U0897585

一城山色半城湖

济南泉·城文化景观特展

济南市文化和旅游局 编

济南出版社

指导单位：

中国文化遗产研究院
山东省文化和旅游厅

主办单位：

济南市文化和旅游局
山东博物馆

承办单位：

济南市博物馆
济南市考古研究院

协办单位：

山东大学考古学院
山东大学博物馆
山东省文物考古研究院
山东省图书馆
章丘区博物馆
长清区博物馆
历城区博物馆
济阳区博物馆
平阴县博物馆
城子崖遗址博物馆

Guiding Units:

China Academy of Cultural Heritage
Shandong Provincial Department of Culture and Tourism

Organizing Units:

Jinan Municipal Bureau of Culture and Tourism
Shandong Museum

Undertaking Units:

Jinan City Museum
Jinan Institute of Archaeology

Co-organizers:

School of Archaeology at Shandong University
Shandong University Museum
Shandong Institute of Cultural Relics and Archaeology
Shandong Library
Zhangqiu District Museum
Changqing District Museum
Licheng District Museum
Jiyang District Museum
Pingyin County Museum
Chengziya Site Museum

编委会

EDITORIAL BOARD

序言

PREFACE

济南是一座因泉而生的城市，泉水是济南具有独特性的景观资源和城市标识。以泉水为源头的“泺”，就见于中国现存最早的文字甲骨文，春秋时期，先民筑建城池，这里成为齐西重镇历下邑。西晋永嘉六年（312），济南郡治由东平陵城移此。千百年来，济南人与泉水相依相伴，演绎了“泉润万物”的人地互动景观，一代又一代的济南人“枕泉而居”，传承着与泉水共生的生活方式，将具有东方特色的泉水文化认知与记忆融入生活的点点滴滴，使这座因泉而生的城市，不仅以独特的泉水景观吸引着世界的目光，更以深厚的文化底蕴讲述着人与自然和谐共生的动人故事。这一文化景观，也是现代化、城市化进程迅猛发展中所剩不多的“大型泉水文化聚落孤本”。

正是基于对“济南泉·城文化景观”独特价值以及城市核心竞争力的深刻认识，济南市委、市政府深入贯彻落实习近平总书记关于加强文化和自然遗产保护传承利用工作重要指示精神，高度重视“济南泉·城文化景观”的保护和申报世界遗产工作，通过编制保护管理规划、实施遗产保护展示工程、举办学术研讨和中外对话交流等多项举措，加快推进申遗工作。2024年，“济南泉·城文化景观”列入更新后的《中国世界文化遗产预备名单》，申遗工作取得阶段性成果。

为彰显济南这座历史文化名城的丰富内涵和深厚底蕴，诠释“济南泉·城文化景观”申遗精髓，加深公众对济南泉水文化、古城风貌的理解与认识，济南市文化和旅游局策划推出“一城山色半城湖——济南泉·城文化景观特展”，并出版本图录。

泉，如生命之脉，潺潺流淌于济南的肌理之中，不仅滋养了这片土地，更赋予了这座城市灵魂与韵律。城，则如同一位温婉的诗人，以千年的时光为笔，以泉水为墨，缓缓诠释着那份独属于济南的泉水文化。让这本图录带您触摸泉城的温度，聆听水流深处文明的回声。

2025年5月

Jinan is a city born of springs, and springs are Jinan's unique landscape resources and an iconic urban symbol. Taking spring water as the source, "Luo" can be found in the oracle bone scripts, the earliest existing scripts in China. During the Spring and Autumn Period, the ancients built a city here, and it became County of Lixia, a strategic stronghold guarding the western frontier of the State of Qi. In the sixth year of Yongjia Era during the Western Jin Dynasty(312), the administrative seat of Jinan Prefecture was relocated from Dongpingling City to this location. For thousands of years, the people of Jinan have been closely associated with spring water, presenting the human-environment interaction landscape of "Spring nourishes all things". Generation after generation of Jinan residents have "dwelled by the springs", maintaining a lifestyle of coexisting with the springs, integrating the understanding and memories of Eastern spring culture into daily life. The city, born from the springs, not only captivates the world with its unique spring water landscapes, but also narrates a moving story of harmonious coexistence between humanity and nature through its profound cultural heritage. This cultural landscape is also one of the few remaining "large spring cultural settlement" in the rapid development of modernization and urbanization.

Successive the Municipal Party Committees and the Municipal Governments of Jinan, grounded in a profound understanding of the unique value inherent in the "Jinan Spring-City Cultural Landscape" and its role as a core urban competitive asset, have thoroughly implemented Xi Jinping's important instructions on strengthening the protection, inheritance, and utilization of cultural and natural heritage. Prioritizing the conservation of the "Jinan Spring-City Cultural Landscape" and inclusion of this spring-nourished cultural landscape into the list of world cultural heritages, they have accelerated progress through formulating protection and management plans, carrying out heritage conservation and exhibition projects, as well as holding academic seminars and dialogues and exchange meetings between China and foreign countries. The "Jinan Spring-City Cultural Landscape" was included in the updated *China's Tentative List of World Cultural Heritage* in 2024, and the application for the inclusion in the list of World Cultural Heritage has achieved phased results.

To showcase the rich connotations and profound historical heritage of Jinan as a renowned historical and cultural city, to interpret the essence of the "Jinan Spring-City Cultural Landscape" World Heritage application, and to deepen public understanding of Jinan's spring culture and the ancient city's historical features, the Jinan Municipal Bureau of Culture and Tourism has curated and launched the special exhibition "A City With Green Mountains and Half A City Surrounded by Lakes: A Special Exhibition on Jinan Spring-City Cultural Landscape", accompanied by the publication of this illustrated catalog.

Springs, like the veins of life, gurgle in the texture of Jinan, which not only nourish the land but also endow the city with soul and rhythm. The city is like a gentle poet, using thousands of years of time as his pen and spring water as his ink, slowly interpreting the spring culture that belongs to Jinan. Now, you can follow this illustrated catalog to touch the temperature of spring water and listen the echoes of civilization from the deepwater flow silently.

May 2025

目录

CATALOGUE

第二章

济水之南

济南古城的发展演变

目录

泉城印象

第一章 城邑肇始

济南地区早期区域中心的变迁

第三章 城泉共生

人文与自然交融的文化景观

趵突泉

泉城印象

IMPRESSION OF THE CITY OF SPRINGS

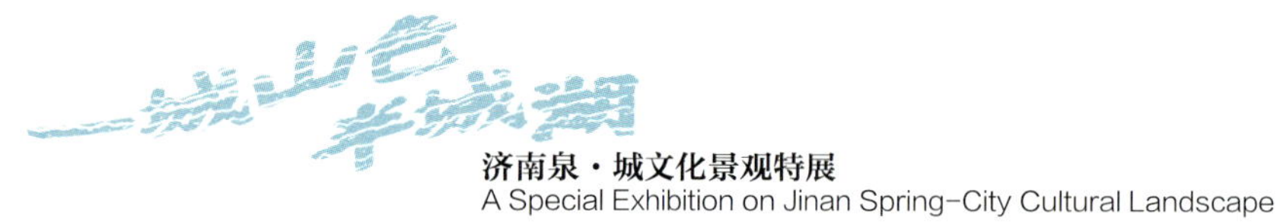

依山面水——济南地形地貌

济南位于黄河下游地区，地处鲁中南低山丘陵与鲁西北冲积平原交接带，是我国北方温带岩溶地貌及岩溶地下水系统广泛分布的区域。

济南以泰山为南部屏障，以古济水（今黄河）两岸为腹地，地势南高北低，呈现由南向北依次为低山丘陵、山前冲积—洪积倾斜平原和黄河冲积平原的地貌形态。

Between Hills and Waters: Geographical Landscape of Jinan

Jinan is located in the lower reaches of the Yellow River, and is located in the transitional zone between low mountainous and hills of south-central Shandong and the alluvial plains of northwestern Shandong. This region represents a key area in northern China where temperate karst landforms and extensive karst groundwater systems are prominently developed.

Taking Mount Tai as its southern natural barrier and the fluvial plains along the ancient Ji River (now the Yellow River) as its central basin, Jinan has the terrain high in the south and low in the north, presenting a landform from south to north that is successively low mountains and hills, piedmont alluvial-proluvial inclined plain, and Yellow River alluvial plains.

济南地理区位示意图
Geographical Location Map of Jinan

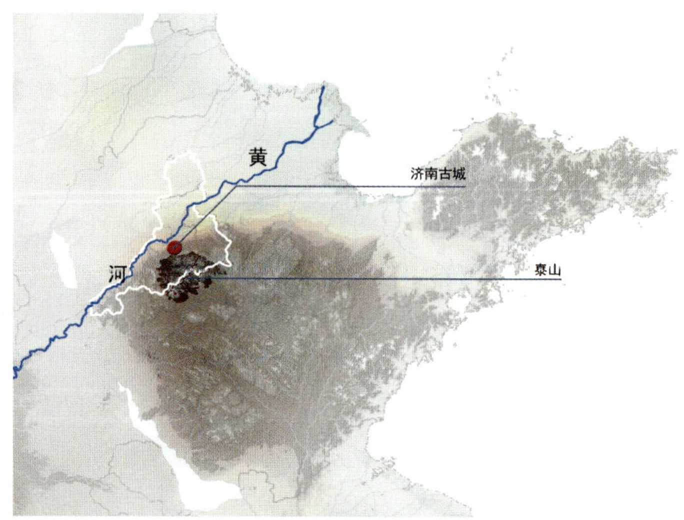

平地涌泉——济南泉水成因

济南以泉水众多而闻名，素有“天下泉城”“世界泉都”的美称。早在宋代，文学家曾巩就评价道：“齐多甘泉，冠于天下。”元代地理学家于钦亦称赞说：“济南山水甲齐鲁，泉甲天下。”济南的泉属于岩溶泉类型，并以泉点众多、水量丰沛、水质优异以及壮观的喷涌景观成为中国岩溶泉的典型代表。济南泉水的形成，是济南地区独特的岩石地层、地质构造和地貌等因素综合作用的结果。

Springs from the Flatlands: Cause of Formation of Jinan's Artesian Springs

Jinan is renowned for its numerous springs, known as the “Spring City of the World” and “World Capital of Springs”. As early as in the Song Dynasty, the literati Zeng Gong praised: “Qi (ancient name for Shandong) abounds in sweet springs, crowned in the world.” Yuan Dynasty geographer Yu Qin also acclaimed: “Jinan’s landscapes supreme in Qilu, and its springs are supreme in the world.” The springs of Jinan belong to the karst springs, distinguished by their numerous outlets, abundant water volume, exceptional water quality, and spectacular gushing scenery, making them the archetypal representation of China’s karst springs. The formation of Jinan’s springs results from the synergistic effects of unique geological features in the region, including distinctive rock stratigraphy, tectonic structures, and topographic characteristics.

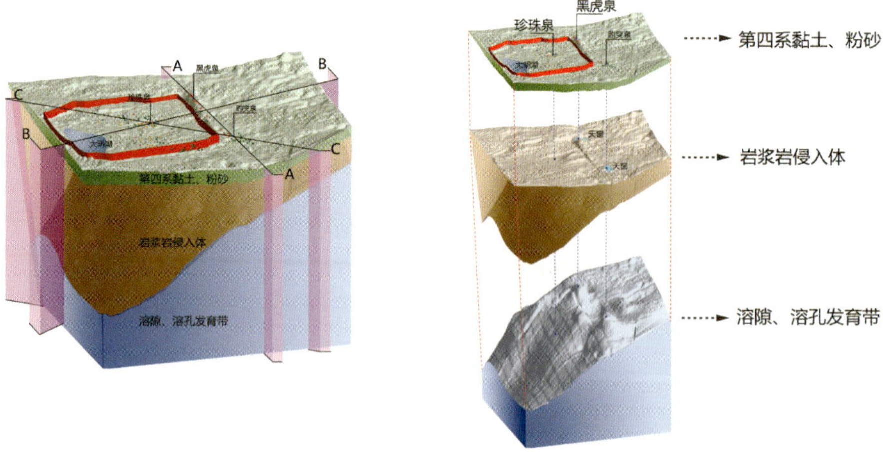

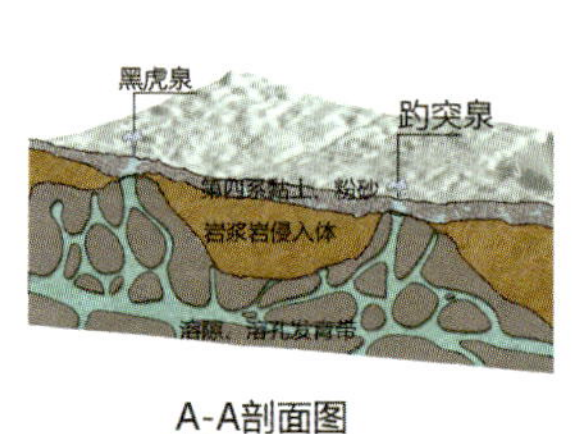

A-A剖面图

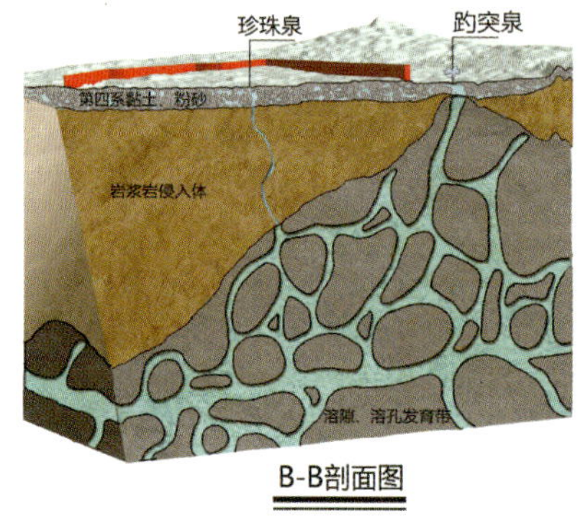

B-B剖面图

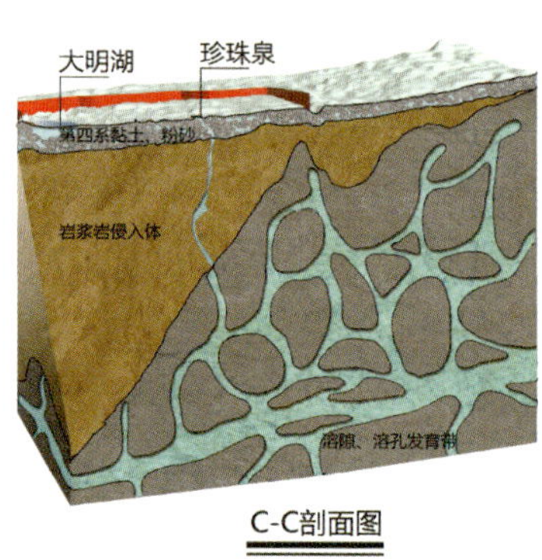

C-C剖面图

千佛山地垒泉水出露区
Spring Water Outcrop Area in Qianfo Mountain Horst

地形地质：济南位于鲁中山地的北缘，地势南高北低，南部山区多为古生界石灰岩，北部则以新生界松散堆积物为主。这种地形和地质结构为地下水的流动和存储提供了良好的条件，雨水和融雪沿着石灰岩的裂隙和孔洞渗透到地下，形成了泉水的直接和间接补给区，最终泉水在沉积层较薄弱处涌出地表，形成天然涌泉。

“三水”转化：大气降水在南部山区形成了大量的山地泉水、地表径流和地下径流，逐步汇集到山前平原区形成多个密集泉群。承压的地下水从这些泉群中涌出，进一步通过天然或人工的河道流向北部平原区或汇集在低洼地带，形成泉水河流、泉水水渠、泉水湿地。最终，这些地表水和地下水通过蒸发冷凝，再次形成大气降水补给到南部山区，至此形成一个相对完整的水循环系统。

集中出露：千佛山地垒泉水集中出露区位于千佛山断裂和羊头峪断裂切割单斜地层形成的千佛山地垒与北部岩浆侵入岩构成三面封闭的构造单元之

上，地质条件特殊，是济南地区岩溶地下水出露最为集中、岩溶水涌量最大的区域，即如今济南古城及周边区域范围，形成了以趵突泉、珍珠泉、金线泉等为代表的众多泉眼。

Topography and Geology: Located on the northern edge of the mountainous areas of the Luzhong region, Jinan has the terrain high in the south and low in the north. The southern mountainous areas are mostly composed of Paleozoic limestones, while the northern areas are dominated by loose deposits from the Cenozoic era. This terrain and geological structure provide favorable conditions for the flow and storage of groundwater. Rainwater and snow-melt infiltrate underground through the cracks and holes of limestone, forming direct and indirect recharge areas for springs. Finally, the weaker parts of the sedimentary layer gush out of the surface, forming natural gushing springs.

Three-Water Transformation : Atmospheric precipitation has formed a large number of mountain springs, surface runoff, and underground runoff in the southern mountainous areas, gradually converging into multiple dense spring clusters in the plain area in front of the mountains. Groundwater under pressure gushes out from these spring clusters and further flows through natural or artificial rivers to the northern plains or gathers in low-lying areas, forming rivers of spring water, spring water channels, and spring water wetlands. In the end, these surface water and groundwater are evaporated and condensed to form atmospheric precipitation again, which is supplied to the southern mountainous areas, thus forming a relatively complete water cycle system.

Concentrated Exposure: The concentrated exposure area of spring water of Qianfo Mountain Horst is located on the Qianfo Mountain Horst formed by the cutting of monocline strata by the Qianfoshan Fault and Yangtouyu Fault, and the three sided closed structural unit composed of northern magma intrusion rocks. With special geological conditions and the most concentrated exposure of karst groundwater and the largest amount of karst water inflow in Jinan, it is now the area where numerous springs represented by Baotu Spring, Pearl Spring, Jinxian Spring emerged in the old city of Jinan and its surrounding areas.

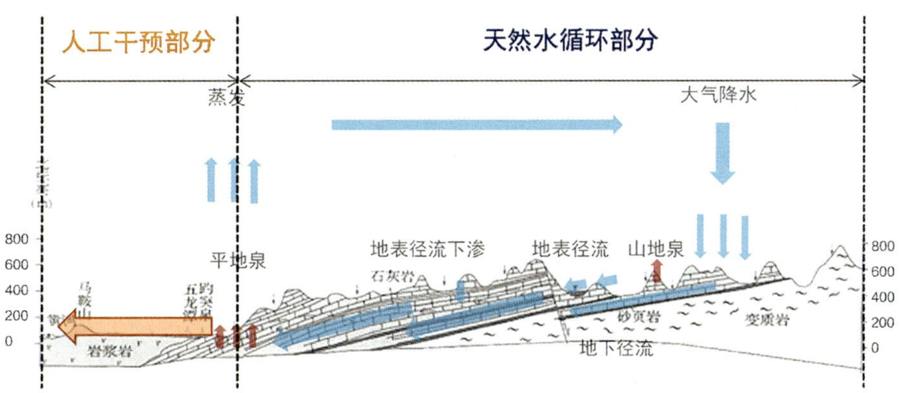

济南岩溶地下水系统“三水”循环示意图
Schematic Diagram of the Three-Water Cycle in the Karst Groundwater System of Jinan

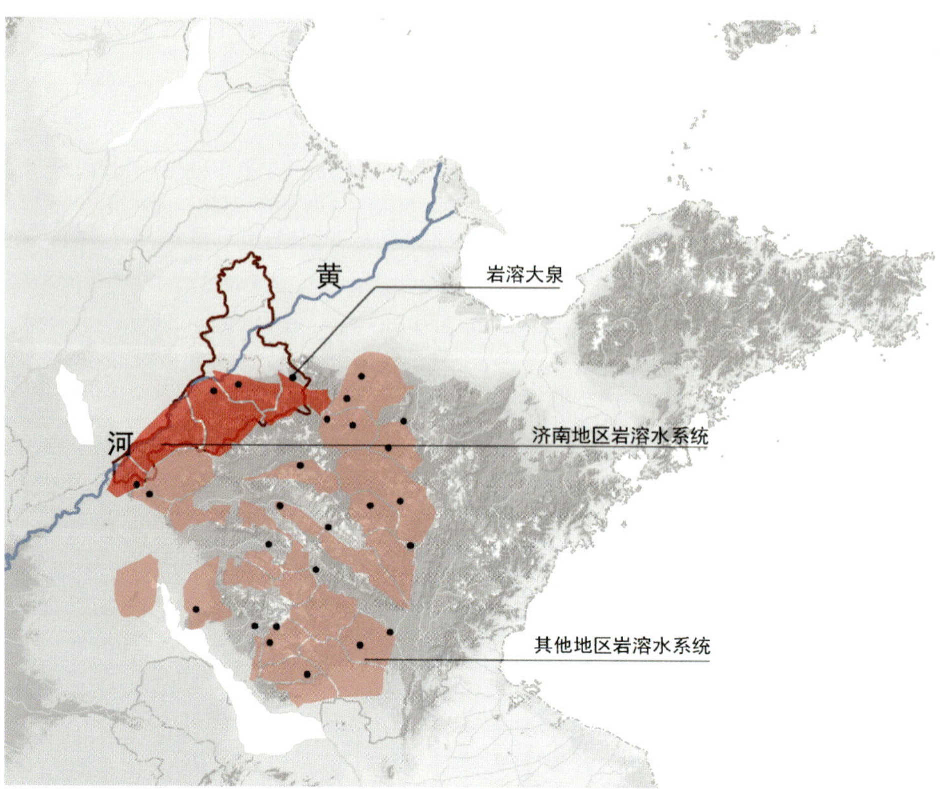

山东地区岩溶水系统分布
Distribution Diagram of Karst Water System in Shandong

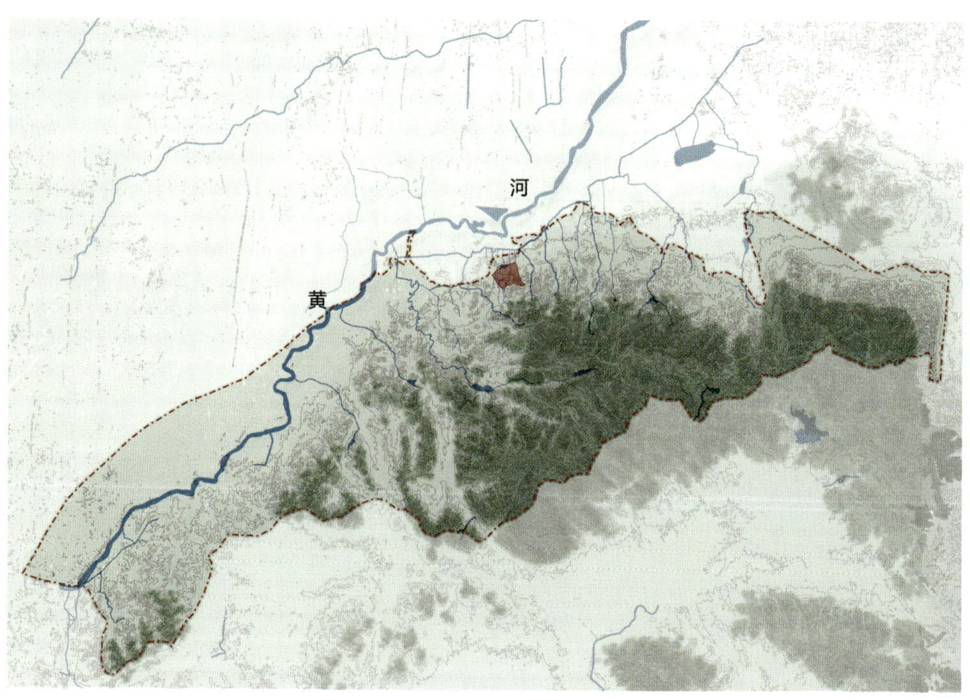

济南岩溶地下水系统范围示意图
Schematic Diagram of the Scope of Jinan Karst Groundwater System

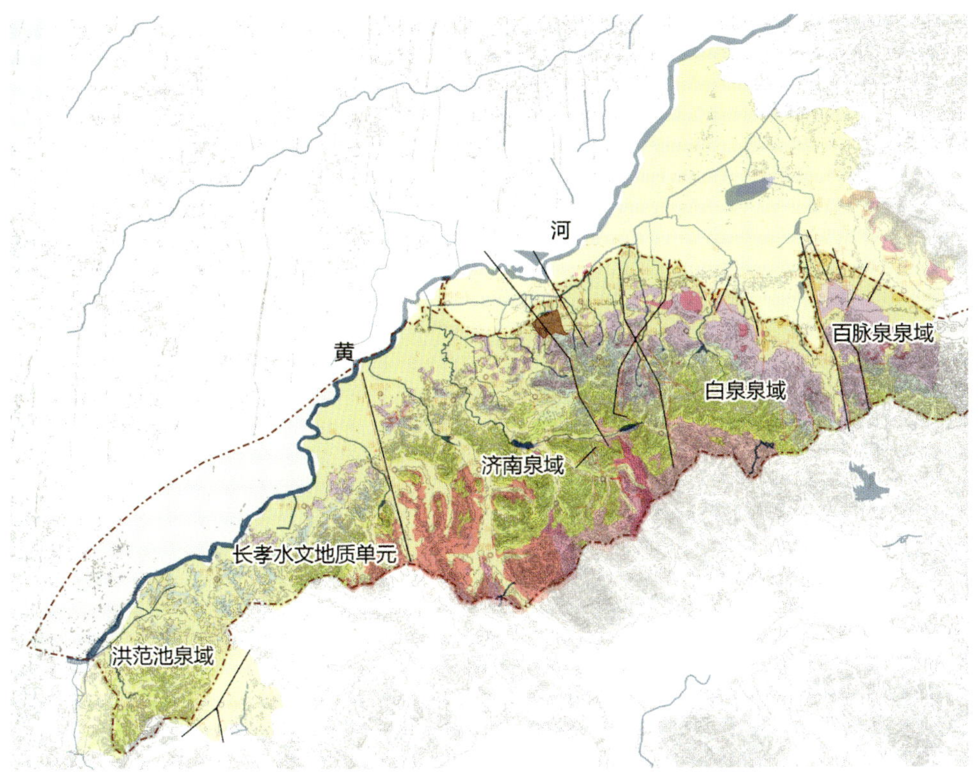

济南岩溶地下水系统地质条件图
Schematic Diagram of Geological Condition of Jinan Karst Groundwater System

千佛山地垒泉水出露区位置示意图
Schematic Diagram of the Location of the Spring Water Outcrop Area in Qianfo Mountain Horst

泉城相融
——“济南泉·城文化景观”

展示古代中国人民对泉水环境治理与利用的高度智慧：“济南泉·城文化景观”以古城冷泉利用系统为核心，将泉水自然环境和城市营建巧妙结合，构建了河道、渠系、蓄水湖泊、水闸、水门等一整套“导—蓄”结合的城市水利系统和以泉水为核心的园林、宅院、寺庙等特色空间。它是集科学、文化、审美于一体，向世人展示古代中国人民理水智慧的重要样本。

体现中国人民尊重自然、寄情山水的精神追求和高超艺术造诣：“济南泉·城文化景观”在长期的演进过程中，形成了丰富多彩的泉水空间，吸引了众多名士前来居住、游赏，催生出独特的泉水评鉴、泉景营造、诗画创作成果。描述过济南名泉的名士超过600人，留下诗篇超过7000首，丰富了中国人民的精神家园。

为当代社会的可持续发展和生态环境保护提供借鉴样本：“济南泉·城文化景观”延续至今，不断演进、优化，在城市化迅猛发展的当代，依然制定并延续着大量尊重自然、保护泉水的规章制度和技术措施，体现了中国从古至今的生态文明思想，是践行绿色中国理念和生态环境保护理念的“大型泉水文化聚落孤本”。

Spring City Symbiosis: “Jinan Spring-City Cultural Landscape”

Demonstration of the High Wisdom of Ancient Chinese People on the Environmental Governance and Utilization of Spring Water: The “Jinan Spring-City Cultural Landscape” takes the utilization system of ancient city cold springs as its core, cleverly combining the natural environment of spring water with urban construction, and constructing a complete set of “channeling and conserving” urban water conservancy systems such as rivers, canal systems, water storage lakes, water gates, and water gates, as well as characteristic spaces such as gardens, courtyards and temples with spring water as the core. It is an important sample of the wisdom integrating science, culture and aesthetics to show the world the wisdom of the ancient Chinese people in water management.

Reflection of the Chinese People’s Respect for Nature, the Spiritual Pursuit of Affection for the Landscape and Superb Artistic Attainments: In the long-term evolution process of “Jinan Spring-City Cultural Landscape”, it has formed a rich and colorful spring space, attracted many celebrities to live and visit, and formed unique achievements in spring evaluation, spring landscape construction, poetry and painting creation. More than 600 celebrities have described Jinan famous springs, leaving more than 7,000 poems, enriching the spiritual home of the Chinese people.

Providing Reference Samples for the Sustainable Development and Ecological Environment Protection in Contemporary Society: The “Jinan Spring-City Cultural Landscape” continues to this day, constantly evolving and optimizing. In the contemporary era of rapid urbanization, a large number of rules, regulations and technical measures for respecting nature and protecting spring water have been established and maintained, reflecting China’s ecological civilization ideas from ancient times to the present. It is one of the few remaining “large spring cultural settlement” that practices the concept of green development in China and ecological environment protection.

第一章 城邑肇始

济南地区早期区域中心的变迁

远在9000年前的新石器时代早期，已有先民在济南地区繁衍生息。四五千年前的大汶口文化和龙山文化时期，济南就已成为鲁北地区重要的区域中心。商代，济南地区成为商王朝经略东方的战略要地。此后，在政权割据时期，济南多位于政权交界地带，是军事门户之地；在和平统一时期，则充当重要城市之间的交通枢纽及区域行政中心、经济重镇的角色。济南东部的章丘一带，河流纵横，水源充足，汉代的东平陵城便选址在此。

Chapter I. The Beginning of the City: The Changes of the Early Regional Centers in Jinan Region

As early as the early Neolithic Period 9,000 years ago, there were already ancients living and multiplying in the Jinan area. During the Dawenkou to Longshan Culture period 4,000-5,000 years ago, Jinan had already become an important regional center in the northern Shandong region. During the Shang Dynasty, the Jinan region had became a strategic location for the Shang Dynasty to manage the East. Afterwards, during the period of political fragmentation, Jinan was mostly located in the border areas of the regime and served as a military gateway. During the period of peaceful reunification, it served as a transportation hub between important cities and a regional administrative and economic hub. The Zhangqiu area in the eastern part of Jinan, crisscrossed by rivers and blessed with abundant water resources, was strategically selected as the site for the Dongpingling City during the Han Dynasty.

文明初曙

济南地区史前文化璀璨夺目。距今 9000 年的后李文化遗存张马屯遗址，揭示了济南最早的人类活动足迹。西河遗址的居住区和房屋以及小荆山遗址的环壕，则展现了新石器时代早期定居聚落的发展。步入大汶口文化时期，章丘焦家城址崛起，它以宏大的规模与显赫的地位，成为济南乃至鲁北地区的区域中心。龙山文化时期，济南史前文明更攀新峰，城子崖古城诉说着古国的辉煌。

济南 9000 年——张马屯遗址

张马屯遗址发现了济南地区最早的人类活动遗迹，将济南的历史推至距今约 9000 年。遗址位于历城区张马屯村南，考古发掘揭示了房址、烧灶等后李文化遗迹，并出土了少量陶片等珍贵实物资料。动植物考古研究表明，张马屯后李文化时期的先民可能已经有栽培粟、黍等作物以及饲养猪等动物的行为。济南的后李文化是目前已知山东境内最早的新石器时代考古学文化。

The Dawn of Civilization

The prehistoric culture in Jinan is dazzling. The Houli Culture site of Zhangmatun, dating back 9,000 years, reveals the earliest footprints of human activities in Jinan. The residential areas and houses of the Xihe Site and the surrounding moats at the Xiaojingshan Site show the development of settlements in the early Neolithic period. Entering the period of Dawenkou Culture, Jiaojia Site in Zhangqiu rose to prominence, becoming the regional center of Jinan and even the northern Shandong region with its grand scale and prominent position. During the Longshan Culture period, the prehistoric civilization of Jinan reached new heights, and the ancient city of Chengziya tells the story of the glory of the ancient country.

Pushing History of Jinan to about 9,000 Years Ago: Zhangmatun Site

Zhangmatun Site is the earliest remains of human activities discovered in Jinan, pushing history of Jinan to about 9,000 years ago. The Site is located in the south of Zhangmatun Village, Licheng District. Archaeological excavations have revealed remnants of Houli Culture, such as house sites and burnt stoves, as well as a small amount of precious physical materials such as pottery shards. Archaeological research on flora and fauna indicates that the ancients of Houli Culture of Zhangmatun may have cultivated crops such as millet and sorghum, as well as raised animals such as pigs.The Houli Culture in Jinan represents the earliest Neolithic archaeological culture in Shandong Province.

房址 F1（东北→西南）
House Site F1 (Northeast → Southwest)

F1 踩踏面（北→南）
Tread surface of House Site F1 (North → South)

F1 室外北部灶址
Kitchen Site in the North outside House Site F1

骨器

Bone Artifact

后李文化时期（距今约 9300—7500）
张马屯遗址出土
济南市考古研究院藏
长 11.35 厘米，宽 3.24 厘米

骨器整体磨制而成，通体扁平呈梯形，下端系有四穿孔等间距分布，上端有磨损使用痕迹。

骨器

Bone Artifact

后李文化时期（距今约 9300—7500）
张马屯遗址出土
济南市考古研究院藏
长 13.7 厘米，宽 2.5 厘米

骨器通体磨光扁平，呈长条状。一端呈斧刃状切削面，另一端逐渐收窄为锥形。

骨簪

Bone Hairpin

后李文化时期（距今约 9300—7500）
张马屯遗址出土
济南市考古研究院藏
长 11.7 厘米，直径 0.52 厘米

兽骨骨壁磨制而成。两端均残。呈长圆锥形，经精细磨光，有使用痕迹。

蚌刀

Mussel Knife

后李文化时期（距今约 9300—7500）
张马屯遗址出土
济南市考古研究院藏
长 7 厘米，宽 3.5 厘米

蚌壳磨制而成，呈舌状。背部和刃部均有磨损痕迹。

居有定所——西河遗址

位于章丘区龙山街道的西河遗址，首次发现了距今8000年左右的后李文化房址。清理的30余座房址保存非常完整，是山东境内新石器时代早中期文化中保存最好、面积最大的一处典型聚落遗存。房址面积多数在30—50平方米，大者70余平方米。有的室内可分为炊煮、加工、储藏、活动和睡眠等功能区。西河先民处于渔猎采集和初级种植饲养相结合的经济形态，过着季节性定居的生活。西河遗址为研究黄河下游地区新石器时代早期考古学文化的面貌特征、年代与分期、经济生活、社会性质及聚落形态等学术课题提供了珍贵的实物资料，入选1997年度全国十大考古新发现。

Residential Settlement: Xihe Site

House sites of Houli Culture dating back to around 8,000 years ago have been discovered for the first time in Xihe Site located in Longshan Street, Zhangqiu District. More than 30 cleared house sites are well-preserved. They are the best preserved and largest remnants of a typical settlement in the early-middle Neolithic period in Shandong Province. Most of the house sites cover an area of 30-50m^2, and area of the largest one is more than 70m^2. Indoor space of some house sites can be divided into cooking, processing, storage, activity, sleeping and other functional areas. The ancients of Xihe lived in an economic form combining fishing, hunting, gathering, and primary planting and breeding, and led a seasonal settled life. Xihe Site has provided precious physical data for the study of academic topics such as the appearance characteristics, chronology and staging, economic life, social nature and settlement forms of the early Neolithic archaeological culture in the lower reaches of the Yellow River. It has been selected as one of the top ten new archaeological discoveries in China in 1997.

*全国重点文物保护单位
National Key Cultural Relics Protection Unit

西河遗址分布范围示意图
Schematic Diagram of the Distribution Range of Xihe Site

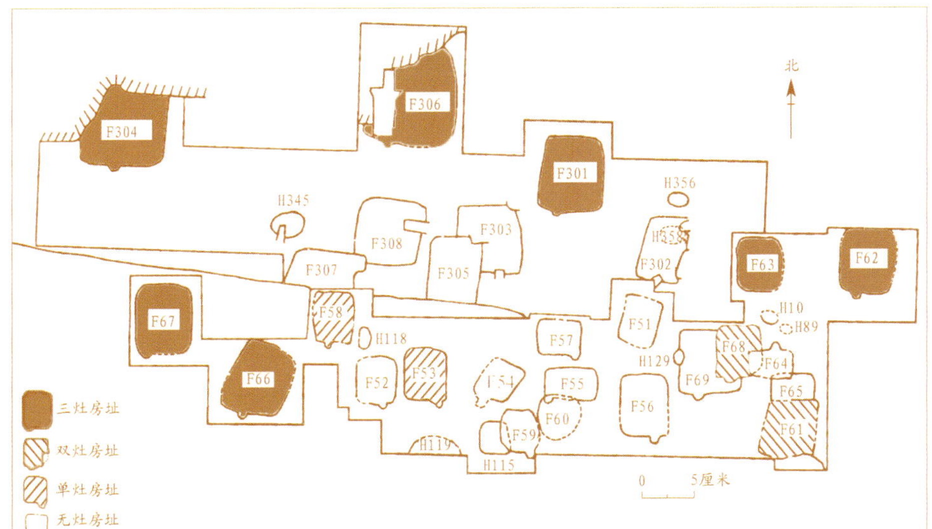

西河遗址居住区主要遗迹平面图
Floor Plan of Main Relics in the Residential Area of Xihe Site

陶兽

Beast-shaped Pottery

后李文化时期（距今约 9300—7500）
西河遗址出土
山东省文物考古研究院藏
面宽 5 厘米，长 7.5 厘米，厚 4 厘米

夹砂陶捏塑而成，兽首呈大圆眼，面部以戳刺纹装饰。

陶釜

Pottery Pot

后李文化时期（距今约 9300—7500）
西河遗址出土
山东省文物考古研究院藏
口径 37.9 厘米，通高 37.7 厘米

环壕聚落——小荆山遗址

小荆山遗址是山东境内目前发现最早、同时期结构最清楚的环壕聚落，位于章丘区刁镇街道茄庄村西南。环壕平面呈圆角三角形，周长约 1130 米的壕沟围绕起面积为 12 万平方米的聚落。环壕内东北部、中部、东南部都分布有房址。遗址发现墓地 3 处，分别位于环壕内、外侧沟沿和环壕外，保存最好的Ⅲ号墓地有 21 座墓葬，分成 3 排，墓主人头向均朝向居住区，是一处有规划的氏族公共墓地。

Circular Trench Settlement: Xiaojingshan Site

Xiaojingshan Site is the earliest circular trench settlement with the most clear structure discovered in Shandong Province, located southwest of Qiezhuang Village, Diaozhen Street, Zhangqiu District, Jinan City. The plane of the circular trench is a rounded triangle, and the trench with a circumference of approximately 1,130m surrounds a settlement with an area of 120,000m^2. There are housing sites distributed in the northeast, central, and southeast parts of the trench. Three burial sites were discovered within the site, located in the circular trench, along the outer edge of in the circular trench, and outside the in the circular trench. The best preserved third burial site consists of 21 tombs, divided into 3 rows. The head of the tomb owner is facing the residential area. It was a planned public cemetery of the clan.

* 全国重点文物保护单位
National Key Cultural Relics Protection Unit

小荆山遗址生活场景复原图
Distribution Diagram of Xiaojingshan Site

石磨盘

Stone Addle-quern

后李文化时期（距今约 9300—7500）
小荆山遗址出土
济南市博物馆藏
长 34.5 厘米，宽 24.3 厘米，高 10.5 厘米

砂岩质，磨盘正面有长期使用的痕迹。

石磨棒

Stone Rollers

后李文化时期（距今约 9300—7500）
小荆山遗址出土
济南市博物馆藏
长 41 厘米，宽 6.5 厘米，厚 7 厘米

石磨盘、石磨棒是人类使用最早、延续时间很长、流传范围很广的谷物加工农具。配合使用的功能大抵有两类，一是去壳，二是磨粉。石磨盘、石磨棒的出土，说明当时原始农业已出现。

陶猪

Pig-shaped Pottery

后李文化时期（距今约 9300—7500）
小荆山遗址出土
济南市博物馆藏
长 12.8 厘米，宽 4.7 厘米，高 6.7 厘米

夹砂红陶，陶猪低首，长吻前突，背部有一穿孔，四肢短小，作行走觅食状。陶猪的整体造型具有野猪风格，造型古朴简单，线条简洁明快，用手捏制而成，是新石器时代早期精彩的陶塑作品。这件陶猪的出现，也表明当时原始部落已经开始饲养家畜。

陶猪

Pig-shaped Pottery

后李文化时期（距今约 9300—7500）
小荆山遗址出土
山东省文物考古研究院
高 5 厘米，长 11.5 厘米

夹砂陶捏塑而成，短腿长嘴，猪身饰戳刺纹。

陶人面

Pottery Human Face

后李文化时期（距今约9300—7500）
小荆山遗址出土
济南市博物馆藏
长3厘米，宽3厘米，厚1.5厘米

夹砂红陶，两道斜直眉，眼窝凹陷，眼眶较高，鼻梁高耸但缺失严重，嘴部不存，整体相对抽象。章丘区小荆山遗址是后李文化时期一处重要遗址，陶器以夹砂红褐陶和红陶为主。

鸟形饰

Bird-shaped Ornament

后李文化时期（距今约9300—7500）
小荆山遗址出土
济南市博物馆藏
长9.5厘米，宽2.3厘米，厚2厘米

通体磨制光滑，一端较宽，钻两孔，下切割一凹槽，逐渐收窄至另一端，形似鸟尾。

骨锥

Bone Awl

后李文化时期（距今约 9300—7500）
小荆山遗址出土
济南市博物馆藏
长 12.8 厘米，宽 3 厘米

顶面较窄，下端较宽，磨制出凿刃，为单面刃。

骨镞

Bone Arrowhead

后李文化时期（距今约 9300—7500）
小荆山遗址出土
济南市博物馆藏
长 7 厘米，宽 1 厘米

骨镞即箭头，是一种常见的狩猎工具。

骨凿

Bone Chisel

后李文化时期（距今约 9300—7500）
小荆山遗址出土
济南市博物馆藏
长 11.3 厘米，宽 5.9 厘米

骨质。通体磨光，圆挺。

史前大城——焦家遗址

位于章丘区龙山街道的焦家遗址，是目前发现的鲁北地区规模最大、等级最高的大汶口文化聚落遗址，也是海岱地区迄今所见年代最早的城址，是中华文明礼制的重要发源地之一，是实证中华五千多年文明史的重要黄河样本。焦家遗址主要为距今5100—4300年的大汶口文化中晚期遗存，目前已发现夯土城墙及其外侧的壕沟，城址面积约12万平方米。城内房址和墓葬布局规律，等级分化十分显著，既揭示了遗址内部文明因素发育的过程，又显示出其高于其他遗址的核心地位，当为同时期济南乃至鲁北地区的中心聚落。焦家遗址的考古发现入选2017年度全国十大考古新发现。

Prehistoric City: The Jiaojia Site

The Jiaojia Site, located in Longshan Street, Zhangqiu District, Jinan City, is currently the largest ruins of the Dawenkou cultural settlement site with the highest level discovered in the northern Shandong region. It is also the earliest city site discovered in the Haidai area to date and is one of the important birthplaces of the ritual system of Chinese civilization. It is an important Yellow River sample for verifying the 5,000 year history of Chinese civilization. The Jiaojia site primarily consists of mid-to-late Dawenkou cultural remains dating back to approximately 5,100-4,300 years ago, with excavated features including rammed-earth walls and surrounding defensive moats, covering a total area of approximately 120,000 m^2. The layout patterns and hierarchical differentiation of housing sites and tombs within the city reveal not only the development process of civilization factors within the site, but also its core status higher than other sites, making it the central settlement of Jinan and even northern Shandong during the same period. The archaeological discoveries of Jiaojia Site had been selected as one of the top ten new archaeological discoveries in China in 2017.

焦家遗址发掘现场
Excavation Site of Jiaojia Site

*全国重点文物保护单位
National Key Cultural Relics Protection Unit

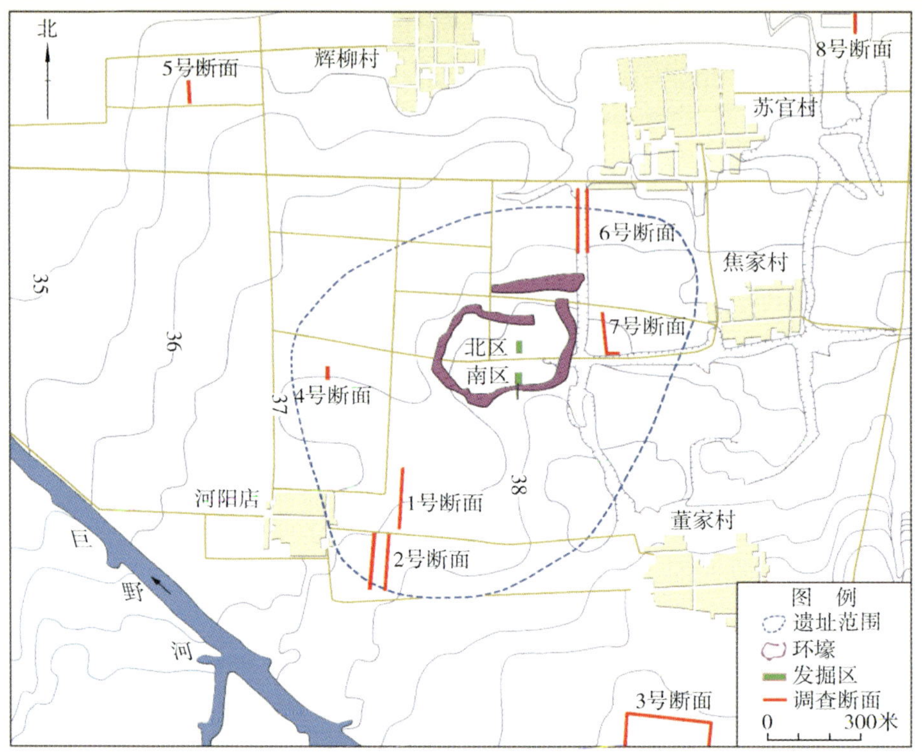

焦家遗址及环壕范围示意图
Schematic Diagram of Jiaojia Site and the Scope of the Circular Trench

祭祀坑
the Sacrificial Pit

陶鬶（guī）

Pottery Gui

大汶口文化时期（距今约 6500—4500）
焦家遗址出土
山东大学博物馆藏
口径 6.6-7.9 厘米，流长 8 厘米，腹径 14.6 厘米，通高 27 厘米，壁厚 0.4—0.5 厘米

夹细云母白陶，一侧有半环状圆形把手，把手下端沿器身饰一周戳印的附加堆纹。手制，经慢轮修整，器身通体磨光。

彩陶壶

Painted Pottery Pot

大汶口文化时期（距今约 6500—4500）
焦家遗址出土
山东大学博物馆藏
口径 7.6 厘米，腹径 13.2 厘米，高 14.2 厘米

泥质红陶。口微敞，尖圆唇，束颈，圆腹，小平底。器表与口沿内侧遍施红陶衣，口沿内外侧各有一条黑彩。泥条盘筑，经慢轮修整，器表磨光。

黑陶高柄杯

Black Pottery Cup with High Handle

大汶口文化时期（距今约 6500—4500）
焦家遗址出土
山东大学博物馆藏
口径 7 厘米，底径 6.9 厘米，高 15.1 厘米，壁厚 0.2—0.3 厘米

泥质黑陶。柄底部饰一周三角形戳印纹。轮制，器身通体磨光。

白陶背壶

White Pottery Back Pot

大汶口文化时期（距今约 6500—4500）
焦家遗址出土
山东大学博物馆藏
口径 12.9 厘米，腹径 22.2 厘米，高 30.3 厘米

夹细云母白陶。敞口，高领，圆肩，鼓腹，两侧附环形耳，另一侧附一平喙突，下腹内收，平底。通体素面磨光。

古国遗踪——城子崖遗址

龙山文化时期的城子崖古国南达泰山北麓，北到刁镇、白云湖一线，东抵长白山西麓，西至小清河支流巨野河。据考古发现，古国已具备“都、邑、聚”三级结构，其中城子崖遗址级别最高，处于金字塔形的顶端，可能已是古国的都城。

城子崖遗址位于章丘区龙山街道龙山村东北，是龙山文化的命名地和代表遗址，也是第一个由中国考古学家发现、发掘并出版考古报告的古遗址。考古发掘证明，这里存在一处包含龙山、岳石、周代三个发展阶段的大型城址，其中龙山文化时期的城子崖古城面积达 20 多万平方米，已经具有早期城市的雏形；岳石文化时期的城子崖则是目前唯一公认的岳石文化城址。城子崖遗址的考古发现入选 1990 年度全国十大考古新发现。

Archaeological Echoes of Ancient states: Chengziya Site

The ancient State of Chengziya in the period of Longshan Culture reached the northern foot of Mount Tai in the south, Diao Town and Baiyun Lake in the north, the western foot of Changbai Mountain in the east, and Juye River, a tributary of Xiaoqing River in the west. According to archaeological discoveries, the ancient state already had a three-level structure of “capital, town and settlement”, among which the Chengziya Site had the highest level, located at the top of the pyramid shape, and may have been the capital of the ancient state.

The Chengziya Site is located in the northeast of Longshan Village, Longshan Street, Zhangqiu District, Jinan City. It is the named site and representative site of Longshan Culture, and also the first ancient site discovered, excavated, and published by Chinese archaeologists with archaeological reports. Archaeological excavations have proven the existence of a large city site including three stages of development, namely Longshan, Yueshi and the Zhou Dynasty. The Chengziya Ancient City during the Longshan Culture period covers an area of over 200,000m^2 and already has the embryonic form of an early city. The Chengziya during the Yueshi Culture period is currently the only recognized Yueshi Culture city site. The archaeological discoveries of Chengziya Site had been selected as one of the top ten new archaeological discoveries in China in 1990.

*全国重点文物保护单位　全国百年百大考古发现

National Key Cultural Relics Protection Unit

Among the list of Top 100 Archaeological Discoveries in China for a Hundred Years

城子崖遗址平面示意图
Schematic Diagram of Chengziya Site

城子崖遗址航拍图
Aerial photo of Chengziya Site

穿孔玉斧

Perforated Jade Axe

龙山文化时期（距今约 4500—4000）
章丘区董家村出土
济南市博物馆藏
长 13.8 厘米，宽 7.6 厘米，厚 1.4 厘米

青玉质，玉质莹润，呈翠绿色，通体磨光。

黑陶镂孔杯

Black Pottery Perforated Cup

龙山文化时期（距今约 4500—4000）
城子崖遗址出土
济南市博物馆藏
口径 9.2 厘米，底径 4.5 厘米，高 14.4 厘米

柄刻镂空，口沿宽且薄，最薄处只有 0.2 毫米，重仅 70 克。蛋壳陶质地细致紧密，表里透黑，不渗水，烧成火候高，表明山东龙山文化制陶业发达，是中国史前制陶工艺的巅峰之作。

鸟喙足鼎

Black Pottery Tripod with Bird Beak-shaped tripod with feet

龙山文化时期（距今约 4500—4000）
城子崖遗址出土
山东省文物考古研究院藏
通宽 27.8 厘米，通高 22.4 厘米

黑陶罍（léi）

Black Pottery Wine Vessel (an ancient urn-shaped wooden wine-vessel)

龙山文化时期（距今约 4500—4000）
城子崖遗址出土
山东省文物考古研究院藏
口径 12.7 厘米，通宽 29 厘米，通高 28.5 厘米

方国都邑

商周时期，济南是中原文明向东拓展的重要枢纽。商代，济南地处商王朝经略东方的战略要冲，出现大辛庄等邑落，展现出东夷文化与殷商文明交流融合的面貌。周初分封，济南地区处于齐、鲁之间，分布着谭、逄等封国。春秋时期，随着齐国崛起，济南成为齐国西南边陲军事重镇，境内泺、历下等城邑见于史册。依托泰山北麓冲积平原与济水航运之利，济南地区农业和工商业取得较大的发展，为秦汉时期历城成为“齐鲁首邑”奠定了基础。

东方重镇（商代）

长清小屯、天桥刘家庄、历城大辛庄等遗址处于东西一线上，这条路线自古以来就是中原地区通向山东半岛的交通要道。济南地区商代遗址的分布及其文化内涵，反映了商王朝对东土的扩张和统治策略，以及商文化和东夷文化交流、融合的图景。

Capital City of a Vassal State

During the Shang and Zhou dynasties, Jinan was an important hub for the eastward expansion of Central Plains civilization. During the Shang Dynasty, Jinan occupied a strategic position in the Shang Dynasty's eastward expansion. Settlements such as Daxinzhuang emerged in the region, reflecting a dynamic cultural exchange and integration between the Dongyi Culture and the Yin-Shang Civilization. During the early Zhou enfeoffment system, Jinan was situated between the Qi and Lu states, with ancient polities such as Tan and Pang distributed across the region. During the Spring and Autumn Period, with the rise of the State of Qi, Jinan became a military stronghold on the southwestern border of Qi. Leveraging the alluvial plains at the northern foothills of Mount Tai and the transport advantages of the Ji River, the Jinan region experienced significant advancements in agriculture, commerce, and industry. These developments laid the foundation for Licheng to emerge as the "preeminent city of Qi and Lu" during the Qin and Han dynasties.

Strategic Stronghold of the Shang Dynasty in the East

Changqing Xiaotun, Tianqiao Liujiazhuang, Licheng Daxinzhuang and other sites are located on the east-west line, which has been a transportation artery from the Central Plains region to the Shandong Peninsula sincc ancient times. The distribution and cultural connotations of Shang Dynasty sites in Jinan reflect the expansion and ruling strategies of the Shang Dynasty over the Eastern territories, as well as the exchange and integration of Shang culture and Dongyi culture.

变形兽面纹亚醜铜罍

Metamorphic Beast Face Pattern "Yachou" Bronze Wine-vessel

商（约前 1600—前 1046）
济南市博物馆藏
通高 42.2 厘米，腹围 87.4 厘米

此器盖内及口内同铭，释为“亚醜”。有“亚醜”铭记的青铜器，宋以来屡有著录，其出土地点明确者多系山东青州苏埠屯。这些器物的出土对于研究商王朝经略东方，以及与东夷的关系有重要意义。

商夷融合——大辛庄遗址

大辛庄遗址是商王朝经略东方的重要据点，年代约在距今 3400—3000 年之间，位于历城区王舍人街道大辛庄村东南。大辛庄遗址在商代是一个集居址、手工业作坊、礼仪中心和墓地于一体的大型中心邑落。大辛庄甲骨卜辞是自 1899 年甲骨文发现以来，首次在商代都城以外地区发现的商代卜辞，表明该遗址在黄河下游地区非同一般的中心地位。大辛庄遗址反映了海岱地区商夷融合的图景，是中华文明多元一体格局进程的具体例证。大辛庄遗址的考古发现入选 2010 年度全国十大考古新发现。

Shang-Yi Cultural Integration: Daxinzhuang Site

Daxinzhuang Site is an important stronghold of the Shang Dynasty’s eastern strategy, dating back to approximately 3,400-3,000 years ago. It is located in the southeast of Daxinzhuang Village, Wangsheren Street, Licheng District, Jinan City. Daxinzhuang Site was a large central settlement in the Shang Dynasty, integrating residential sites, handicraft workshops, ceremonial centers and cemeteries. Oracle bone inscriptions at Daxinzhuang are Shang Dynasty inscriptions discovered for the first time outside the capital city of the Shang Dynasty since the discovery of inscriptions on oracle bones in 1899, indicating that the site holds an extraordinary central position in the lower reaches of the Yellow River. Daxinzhuang site epitomizes the cultural synthesis between Shang civilization and the indigenous Yi peoples in the Haidai region (Shandong Peninsula), serving as a material testament to the historical processes shaping the ’pluralistic unity’ framework of Chinese civilization. The archaeological discoveries of Daxinzhuang Site had been selected as one of the top ten new archaeological discoveries in China in 2010.

*全国重点文物保护单位
National Key Cultural Relics Protection Unit

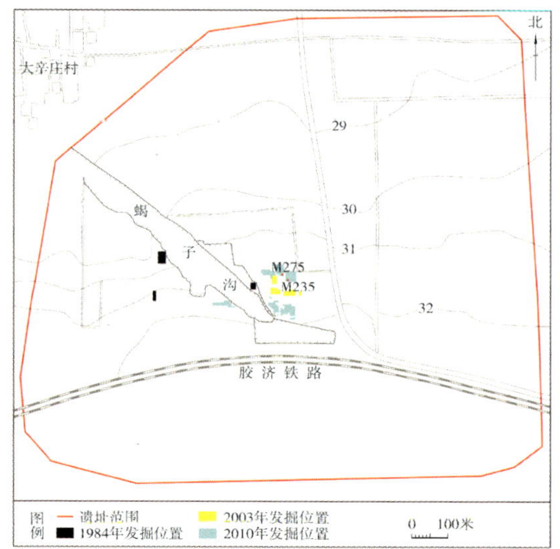

大辛庄遗址历年发掘位置示意图
Schematic Diagram of the Excavation area of Daxinzhuang Site over the Years

大辛庄遗址发掘现场
Excavation Site of Daxinzhuang Site

大辛庄遗址发现的商代墓葬
Shang Dynasty Tombs Discovered at the Daxinzhuang Site

甲骨卜辞

大辛庄卜甲是在安阳殷墟和郑州商城以外第一次发现的商代有字卜甲。经缀合的一版龟腹甲，其卜辞内容主要记“御四母”以御除灾殃、沐浴洁身以祭，以及卜问神祇是否徙降受享诸事，占卜主体应是邑落内的上层权贵，卜甲距今已有3200年的历史。

Inscriptions on Tortoise

Inscriptions on tortoise at Daxinzhuang represent the discovery for the first time of Shang Dynasty inscribed divination bones outside the core Shang capitals of the Yin Ruins in Anyang and Zhengzhou Shang Dynasty Site. A reconstructed turtle plastron (carapace) from the Shang dynasty contains divinatory inscriptions documenting ritual protocols to "appease the Four Ancestress Deities" for calamity mitigation, purificatory ablutions before sacrifices, and divinations inquiring whether deities would descend to receive offerings. The divination patron was likely elite nobility within a walled settlement. This oracle bone dates back approximately 3,200 years.

大辛庄遗址出土商代卜甲
Oracle Tortoise Shell of the Shang Dynasty unearthed at Daxinzhuang Site

卜甲
Oracle Tortoise Shell
商（约前 1600—前 1046）
大辛庄遗址出土
济南市考古研究院藏
长 17.4 厘米，宽 10.2 厘米

龟腹甲。背面钻凿沿中缝排列有序，经灼烧，正面兆痕清晰。

卜骨
Oracle Bone
商（约前 1600—前 1046）
大辛庄遗址出土
济南市考古研究院藏
长 29.5 厘米，上宽 6.5 厘米，下宽 16.5 厘米

肩胛骨，经人工锯截而成。背面钻凿经灼烧，正面有细小兆纹和刮削划痕。

大辛庄墓地所见族徽

大辛庄商代墓葬出土青铜器上发现了多种不同的族徽，表明在大辛庄及其周边生活着多个商代家族，来自不同地区或族群的官僚和军队聚集于大辛庄，为商王朝经略东方的战略而服务。特别是“索”字铭文铜器，为研究商末征夷方路线提供了新的资料。

Clan Emblem Discovered at the Daxinzhuang Tomb

Various ethnic emblems were discovered on bronze artifacts unearthed from the Shang Dynasty tombs in Daxinzhuang, indicating that multiple Shang Dynasty families living in and around Daxinzhuang. Bureaucratic officials and armies from different regions or ethnic groups gathered in Daxinzhuang to serve the Shang Dynasty's strategy of managing the East. Bronze vessels bearing the “Suo” clan insignia particularly provide new epigraphic materials for reconstructing the military routes of the late Shang campaigns against the Yi tribes.

铜爵 M275：2 鋬（pàn）内腹外壁有族徽铭文“▼”。此族徽目前发现很少。

Bronze Jue M275：2 features the clan emblem “ ▼ ” inscribed on the exterior wall beneath the handle. This clan emblem has been rarely documented in extant archaeological records.

两件带有“ ”字族徽铭文的铜爵出自同一墓地中的两座墓葬。该族徽从辛从又，可隶定为“𰃮”字，不见于以往的铜器著录，为大辛庄商代家族的存在提供了直接的文字证据。

Two bronze jue vessels inscribed with the “ ” clan insignia were unearthed from two separate tombs within the same cemetery. This clan insignia, combining the glyphs “ 辛 ” and “ 又 ”, can be transcribed as the character “𰃮”. As it is not attested in previous bronze records, it provides direct epigraphic evidence for the existence of a Shang-dynasty lineage associated with the Daxinzhuang site.

铜鼎 M275：14 内底有一族徽铭文“口”，该族可能为改（jǐ）姓或妊（rèn）姓族氏。

Bronze Tripod M275：14 bears the clan emblem “口” on its interior base, indicating that clan may be a clan with the surname Ji or the surname of Ren.

铜矛

Bronze Spear

商（约前 1600—前 1046）
大辛庄遗址出土
济南市考古研究院藏
长 17.2 厘米，宽 5.3 厘米，厚 0.25—2.8 厘米

叶部整体近等腰三角形，叶肩弧形，前锋尖锐，中部四棱锥状脊直贯矛锋，双叶有刃。

有阑兽面纹曲内戈

Flanged Dagger-Axe (Ge) with Beast-Face Pattern and Curved Tang

商（约前 1600—前 1046）
大辛庄遗址出土
济南市博物馆藏
通长 27 厘米，援长 17.5 厘米，最宽 4.9 厘米，厚 0.9 厘米

窄长条三角形援，前锋尖锐，援部扁平较薄，中脊略显，有上下阑。曲内弧尾。

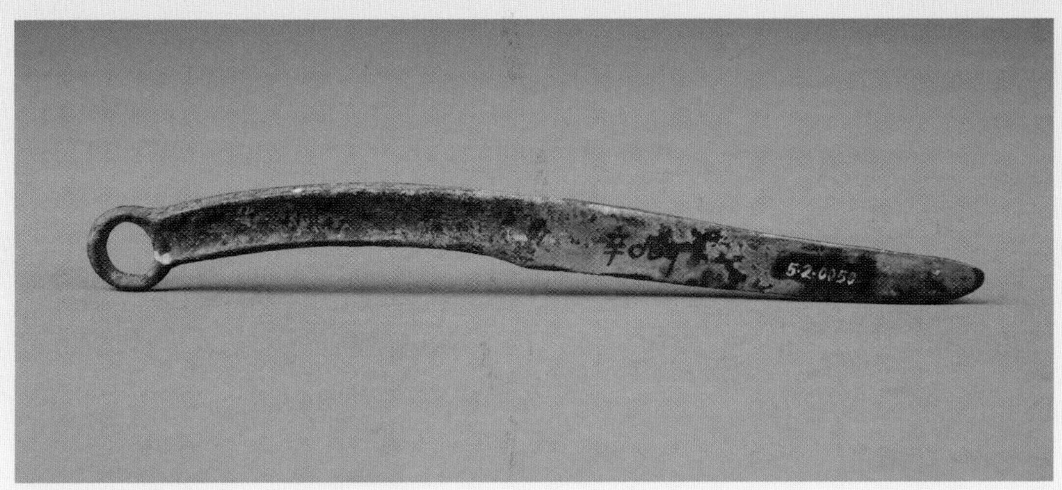

凹柄环首铜削

Bronze Knife with Concave Hilt and Ring Pommel

商（约前 1600—前 1046）
大辛庄遗址出土
济南市博物馆藏
长 23.5 厘米，最宽处 1.9 厘米，厚 0.8 厘米

直背，凹柄，刃较平，刃至顶端向上收敛成锋尖，锋尖不翘起，与背相齐。柄作环首是为了穿绳佩戴。

兽面纹铜削

Bronze Knife with Beast-Face Pattern

商（约前 1600—前 1046）
大辛庄遗址出土
济南市博物馆藏
长 24.5 厘米，最宽处 2.5 厘米，厚 0.4 厘米

平背，平刃，刃至顶端向上收敛成锋尖，锋尖微翘。刀身细长，柄与刃分界明显，交接处呈直角。

玉戚

Yu Qi (Axe with a Jade Handle or Jade Ornament)

商（约前 1600—前 1046）
大辛庄遗址出土
济南市考古研究院藏
长 10 厘米，宽 6.49 厘米

蛇纹石质。灰青色，表面褐色沁斑。近长方形，弧顶，刃角略侈，双面刃略弧。肩部两侧一组扉棱。钺身中部对钻一圆孔。

玉圭

Jade Gui (Elongated Pointed Tablet of Jade)

商（约前 1600—前 1046）
大辛庄遗址出土
济南市考古研究院藏
长 18.2 厘米，宽 5.9 厘米

近长方形，白色，通体抛光，平首微沁褐色，弧形双面刃。

玉镞

Jade Arrowhead

商（约前 1600—前 1046）
大辛庄遗址出土
济南市考古研究院藏
长 7 厘米，最宽处 1.4 厘米，厚 0.29—0.5 厘米

扁长圆锥形，通体抛光，脊部微高，两侧磨成刃，铤、脊相连无明显分界，铤部呈尖锥状。

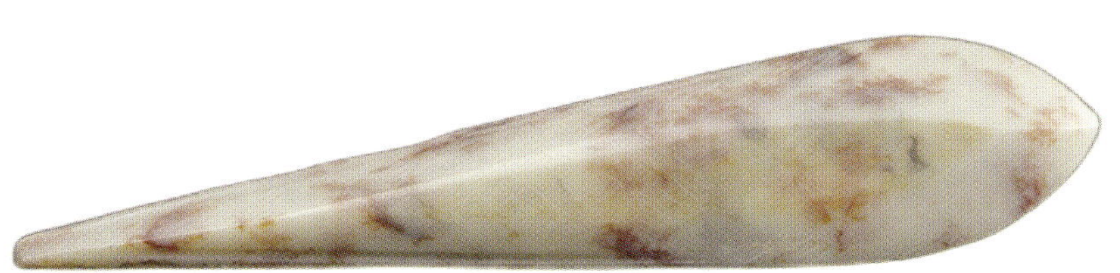

玉矛

Jade Spear

商（约前 1600—前 1046）
大辛庄遗址出土
济南市考古研究院藏
长 10.4 厘米，宽 5.9 厘米，厚 0.47—0.87 厘米

透闪石软玉，整体呈不规则三角形，又似叶形。矛为商周时期常用的兵器，但玉矛一般作为礼器使用，象征身份和权力。

玉兽

Jade Beast

商（约前 1600—前 1046）
大辛庄遗址出土
济南市考古研究院藏
长 5.2 厘米，宽 2.9 厘米

青白色，有褐色沁斑。薄体，马回头状。双腿跪地，尾巴直伸。

骨笄

Bone Hairpin

商（约前 1600—前 1046）
大辛庄遗址出土
济南市考古研究院藏
长 8.09 厘米，最宽 2.13 厘米，厚 0.19—0.76 厘米

骨制。磨光。簪首为一站立凤鸟。锯齿状长冠向后高耸卷曲，尾部下垂。足部呈梯形立于簪体上方。目视前方，神态安详。尾端尖长锋利。凤身饰三条弦纹。

兽面纹面具
Mask with Beast Face Pattern
商（约前 1600—前 1046）
大辛庄遗址出土
济南市博物馆藏
厚 3 厘米

面具作牛头状，威严神武。大辛庄遗址出土的这件铜面具，很可能与军事祭祀有关。

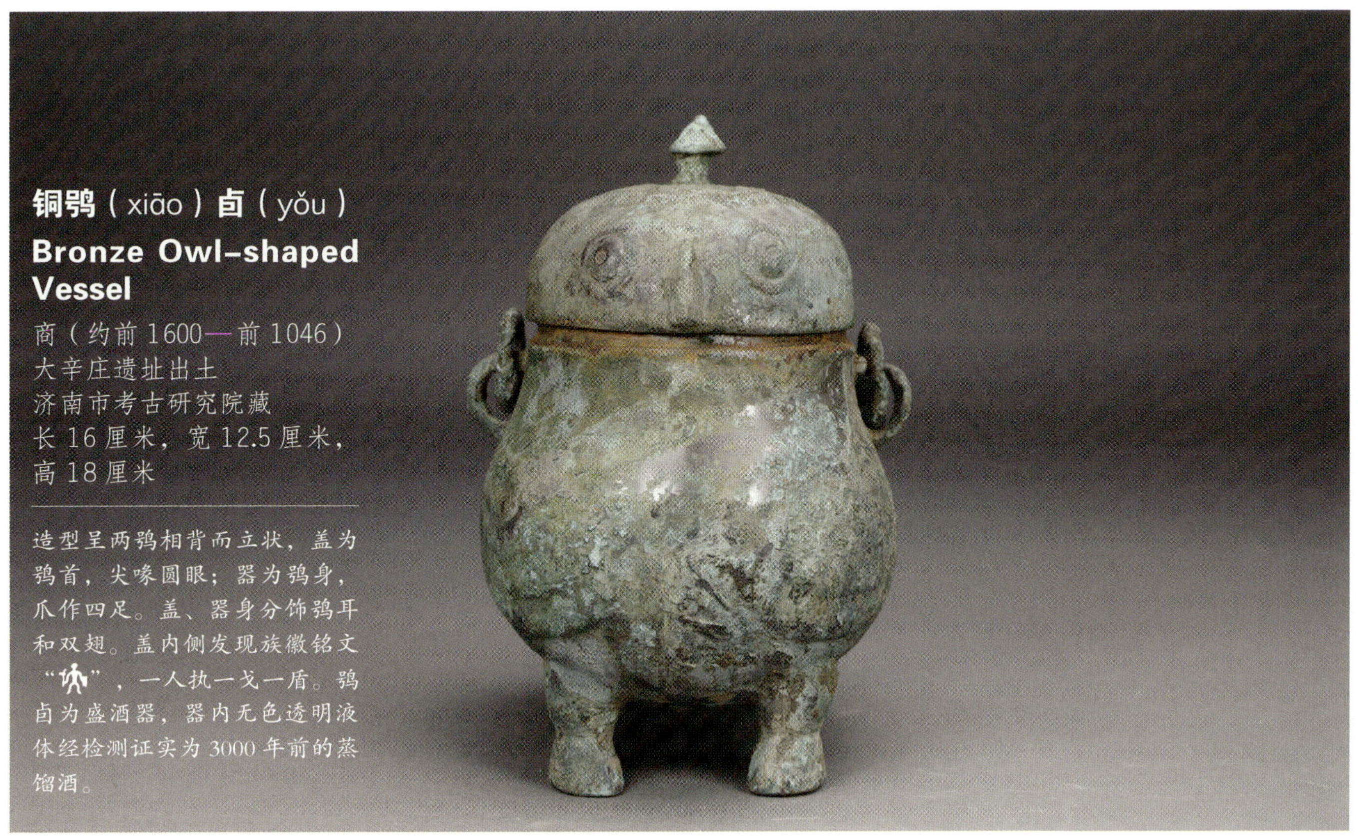

铜鸮（xiāo）卣（yǒu）
Bronze Owl-shaped Vessel
商（约前 1600—前 1046）
大辛庄遗址出土
济南市考古研究院藏
长 16 厘米，宽 12.5 厘米，高 18 厘米

造型呈两鸮相背而立状，盖为鸮首，尖喙圆眼；器为鸮身，爪作四足。盖、器身分饰鸮耳和双翅。盖内侧发现族徽铭文“[illegible]”，一人执一戈一盾。鸮卣为盛酒器，器内无色透明液体经检测证实为 3000 年前的蒸馏酒。

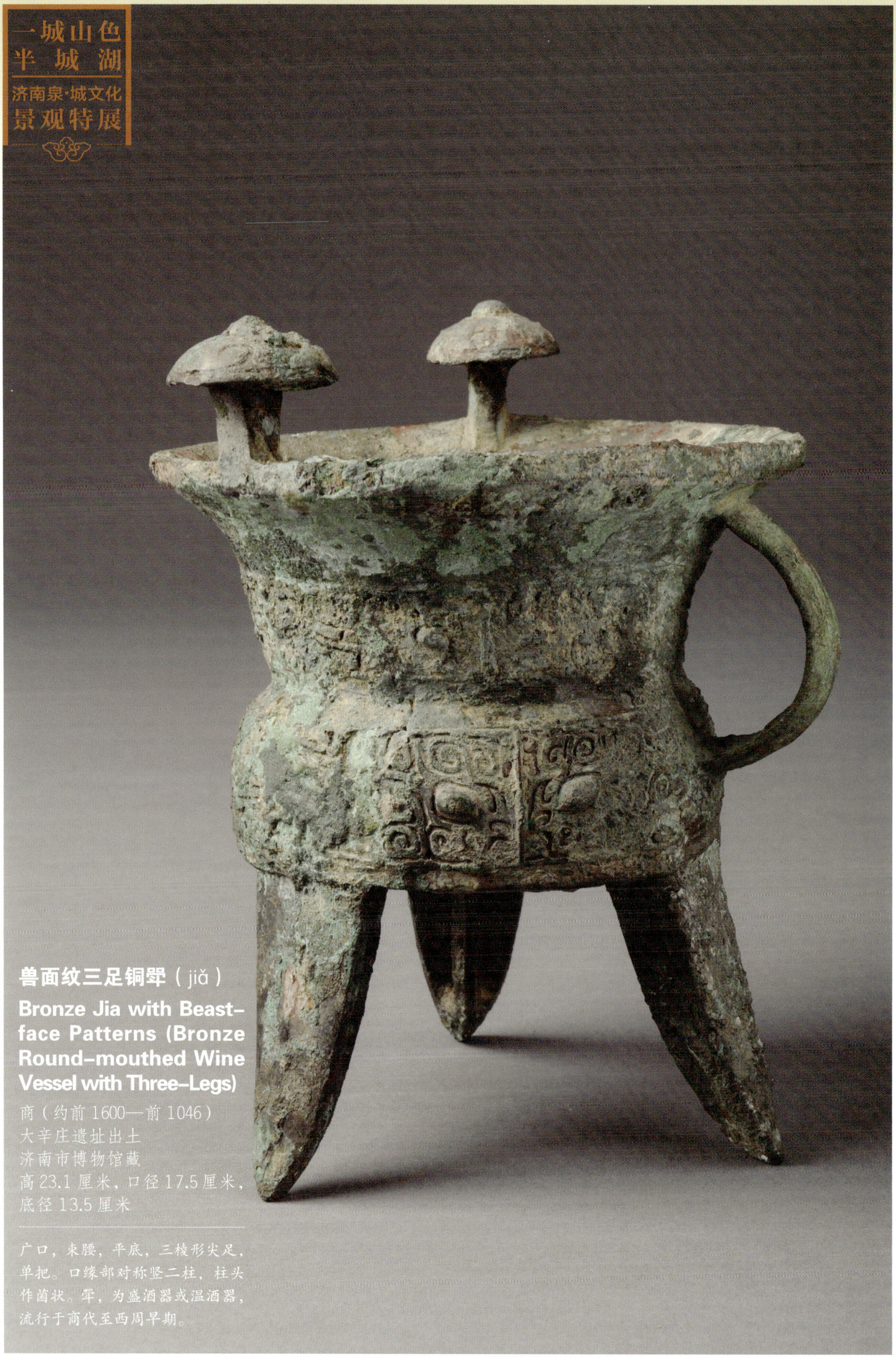

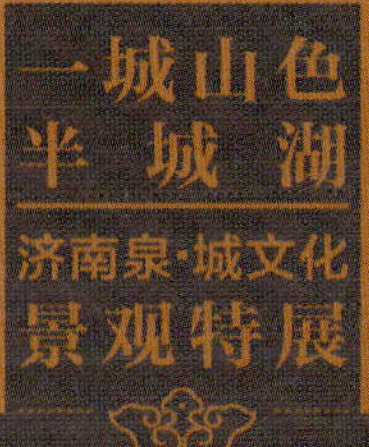

兽面纹三足铜斝（jiǎ）

Bronze Jia with Beast-face Patterns (Bronze Round-mouthed Wine Vessel with Three-Legs)

商（约前 1600—前 1046）
大辛庄遗址出土
济南市博物馆藏
高 23.1 厘米，口径 17.5 厘米，底径 13.5 厘米

广口，束腰，平底，三棱形尖足，单把。口缘部对称竖二柱，柱头作菌状。斝，为盛酒器或温酒器，流行于商代至西周早期。

王族邑落——刘家庄遗址

商末，东临济南古城的刘家庄区域人口稠密，祭祀活动频繁，与大辛庄遗址乃至商都殷墟有着紧密的联系。刘家庄遗址发现了商代窖穴和祭祀坑，少量坑内出土卜骨。刘家庄墓地的青铜器上共发现 7 种族徽铭文。墓地内可大致分辨出该地统治阶层“役”族贵族墓葬区、“子”姓商族“子工”族墓葬区以及平民阶层墓葬区。

Royal Clan Settlement: Liujiazhuang Site

At the end of the Shang Dynasty, the Liujiazhuang area adjacent to the ancient city of Jinan had a dense population and frequent sacrificial activities, which were closely related to the Daxinzhuang site and even Yin Ruins of Shang Dynasty. Shang Dynasty cellars and sacrificial pits were discovered at Liujiazhuang site, with a small number of divination bones unearthed from the pits. A total of seven ethnic emblem inscriptions were discovered on the bronzes in Liujiazhuang Cemetery. Within the cemetery, three distinct burial zones can be roughly identified, i.e., the aristocratic burial section of the “Yi” clan (ruling class), the burial section of the “Zigong” clan (a sub-lineage of the Zi-surnamed Shang elite), and the commoner burial section.

商代祭祀坑出土的完整牛骨
Complete Ox Bones from Sacrificial Pits in the Shang Dynasty

兽面纹铜銎（qióng）内戈

Bronze Socketed-Haft Dagger-Axe with Beast Face Patterns

商（约前 1600　前 1046）
刘家庄遗址出土
济南市博物馆藏
长 25 厘米，宽 6.5 厘米，厚 2.6 厘米

銎内戈。宽援较长，呈长条三角形，有中脊，前锋较锐。

孔雀石鸟

Malachite Bird

商（约前 1600—前 1046）
刘家庄遗址出土
济南市考古研究院藏
长 3.8 厘米，宽 3.0 厘米，厚 0.8 厘米

鸟形，尖嘴，圆眼，圆首，两翼展开，短尾，双翼刻划卷云纹。中部一钻孔，腹下有一长方形凸起。

铜斝（jiǎ）

Bronze Jia

商（约前 1600—前 1046）
刘家庄遗址出土
济南市考古研究院藏
通高 28.1 厘米，通宽 24.9 厘米

侈口，方唇，口上两菌状立柱，束颈，鼓腹，分裆，柱足中空。温酒器，也被用作礼器。

父辛铜觚（gū）

Fu Xin Bronze Gu

商（约前 1600—前 1046）
刘家庄遗址出土
济南市博物馆藏
口径 15.6 厘米，底径 9 厘米，高 27.4 厘米

喇叭形口，细腰，圈足。圈足内壁有铭文“父辛”二字。

乳钉纹铜簋（guǐ）

Nipple-patterned Bronze Goblet

商（约前 1600—前 1046）
刘家庄遗址出土
济南市博物馆藏
口径 19.4 厘米，底径 14 厘米，高 15 厘米

石器，方唇外折，侈口束颈，圆腹外鼓，圈足。器内底铸有一族徽。

赫赫举族——小屯遗址

位于长清区归德街道小屯村东的小屯遗址，出土商代晚期青铜器100余件，其中17件有铭文，7件带有“举”族徽铭文。“举”是商人诸多族群中规模、势力较大的一支，他们随着商人的扩张散布各地，包括今天的济南周边地区。

Prominent Clan: Xiaotun Site

The Xiaotun Site located to the east of Xiaotun Village, Guide Street, Changqing District, has unearthed more than 100 bronze artifacts from the late Shang Dynasty, including 17 with inscriptions and 7 with “ju” ethnic emblem inscriptions. “Ju” was among the largest and most powerful of the many clans of the people of Shang Dynasty. They have spread all over the world with the expansion of the people of Shang Dynasty, including the surrounding areas of Jinan today.

*山东省重点文物保护单位
Key Cultural Relics Protection Unit of Shandong Province

小屯遗址出土青铜鼎“举”字铭文拓片
Rubbings of “ju” Inscriptions on the Bronze Tripod unearthed at Xiaotun Site.

青铜举方鼎

Ju Bronze Square Tripod

商（约前 1600—前 1046）
长清小屯遗址出土
山东博物馆藏

内壁铸有铭文“举祖辛禹”及一亚形族徽。目前发现的商代晚期举族铜器基本都在河南殷墟和山东长清一带。举族在商王室中担任重要官职，受到商王重用，武丁时期曾多次参与征伐西部边邑异族的战争。山东的举族可能是商末征伐东方的一支主力军。长清一带应是山东举族聚居之地。

晚商封地——臧庄商墓

济右走廊南端的平阴，襟带山河，是历史上兵家必争之地。晚商时期，洪范池镇臧庄村一带存在着一个区域政治中心，可能是商人为经略东方而建立的据点。在臧庄村发现了一座商代晚期墓葬，有3人合葬，墓主居中，侧旁两人应为殉人。墓中出土陶罐3件，铜爵、觚、鼎各1件，遗物的表层均有腐朽的编织物痕迹。铜爵把手内腹壁铸有“子义”二字，表明墓主应属“子”姓王族。从墓葬规模及殉人殉物的数量上来看，墓主可能是当时商王兄弟之子，或先王及其兄弟之子，而臧庄就是其封邑所在。

Late Shang Fief: The Shang Tomb at Zangzhuang

Pingyin at the southern end of the Jiyou Corridor, with mountains and rivers in its embrace, was a strategically important location for military strategists in history. In the late Shang Dynasty, there was a regional political center in Zangzhuang Village, Hongfanchi Town, which may have been a stronghold established by the people of the Shang Dynasty for the purpose of managing the East. A late Shang Dynasty tomb was discovered in Zangzhuang Village, where three people were buried together. The tomb owner was in the center, and the two people on the side were likely accompanying burials. Three pottery jars were unearthed from the tomb, along with one bronze noble vessel, one bronze goblet, and one bronze tripod. The surface of the relics showed signs of decaying woven fabrics. The two characters “Ziyi” are cast on the inner abdominal wall of the bronze duke’s handle, indicating that the tomb owner should belong to the royal family with the surname “Zi”. Judging from the scale of the tombs and the number of martyred objects, the owner of the tomb may have been the son of the Shang King’s brothers or the son of the first king and his brothers at that time, and Zangzhuang was where his settlement was located.

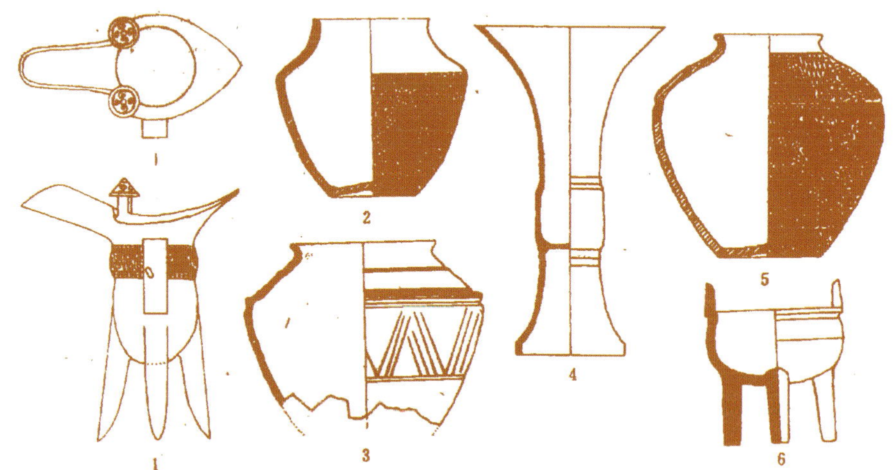

臧庄商墓出土部分器物线稿

Line Diagram of Part Artifacts Unearthed from the Shang Tomb at Zangzhuang

1. 铜爵　2. 陶罐　3. 陶罐　4. 铜觚　5. 陶罐　6. 铜鼎

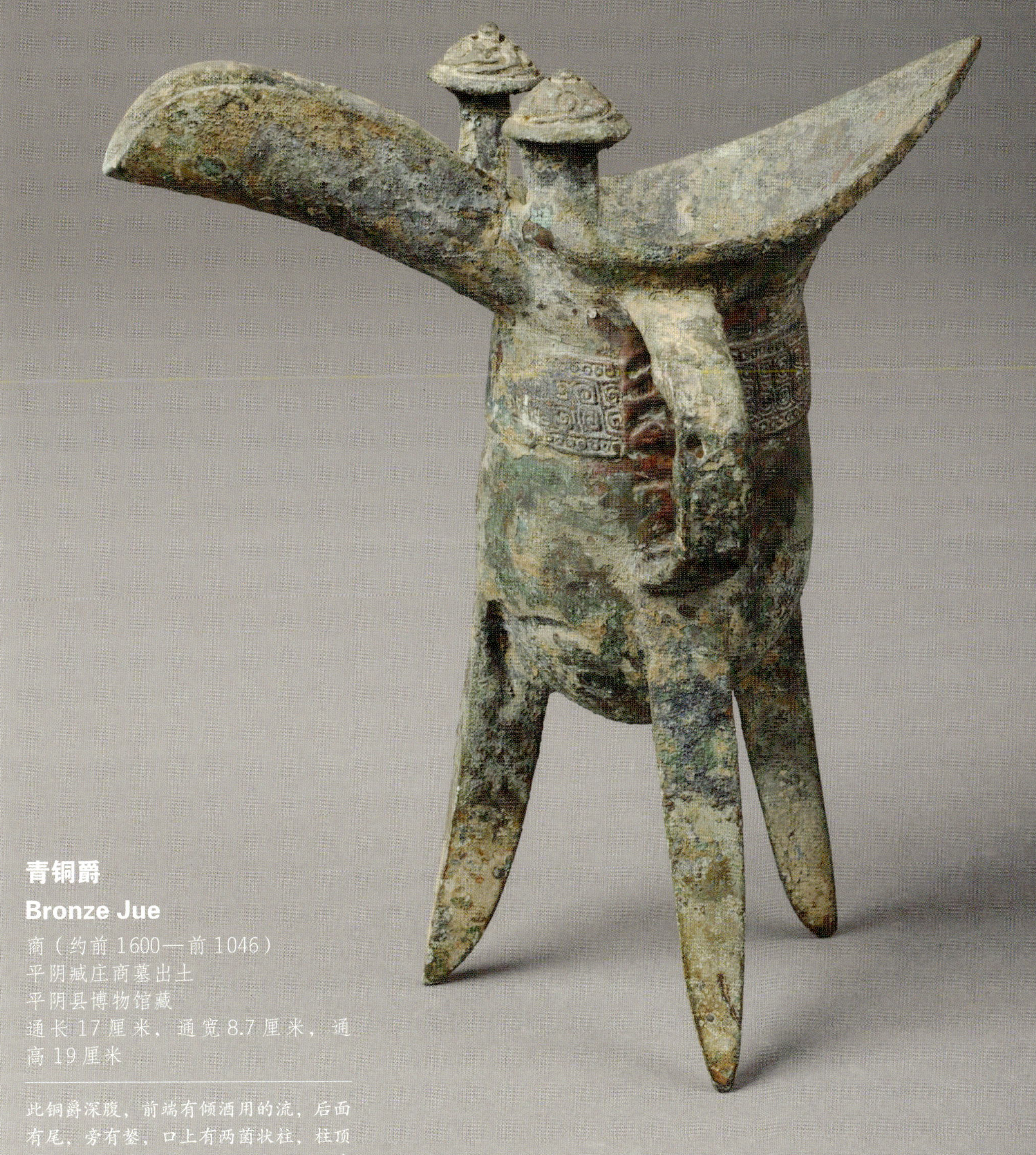

青铜爵

Bronze Jue

商（约前 1600—前 1046）
平阴臧庄商墓出土
平阴县博物馆藏
通长 17 厘米，通宽 8.7 厘米，通高 19 厘米

此铜爵深腹，前端有倾酒用的流，后面有尾，旁有鋬，口上有两菌状柱，柱顶饰漩涡纹，爵底有 3 个尖的高足。铜爵腰部有一周云雷纹，鋬侧的爵外壁铸有“子义”二字。

封国林立（周代）

周道向东最远通达齐都临淄，是周王朝联系山东境内封国的重要交通线。济南境内的周道就建在济右走廊之上，走廊上的谭、祝、邿（shī）、宿、须句（qú）、卢等国也因周道而同周都洛邑紧密地联系在一起。春秋战国时代，济南属齐国，称“泺”“鞍”“历下”等邑，为齐国西陲重镇。

Ancient States in Great Numbers in the Zhou Dynasty

The Zhou Royal Road extended eastward to its farthest terminus at Linzi, the capital of the Qi State, forming a crucial transportation artery that connected the Zhou dynasty court with its feudal states in Shandong. The Zhou Royal Road in Jinan was built on the Jiyou Corridor, and the states of Tan, Zhu, Gui, Su, Xuju, and Lu along the corridor were closely connected to the capital city of Luoyi due to the Zhou Road. During the Spring and Autumn Period and the Warring States Period, Jinan was divided into the State of Qi and was known as a major town in the western border of Qi, with cities such as “Luo”, “An” and “Lixia”.

周代济南地区主要封国
Major Ancient States in Jinan during the Zhou Dynasty

国名	国君姓	地望	备注
谭国	子	章丘平陵城一带	前 684 年为齐所灭
祝国	妊	今槐荫区古城村	前 768 年为齐所灭
邿国	妫	长清	前 560 年为鲁所灭
逄国	姜	初在近临淄一带，后迁至今济南济阳区一带	
卢国	姜	长清归德街道	齐地，后改邑
甗国	姜	今济南西南	齐地
著国		今济南济阳区	齐地
崔国	姜	今济南章丘区西北土城村	前 546 年除

逄国贵族——刘台子逄国贵族墓地

相传炎帝后裔有人名陵，商初受封于逄（páng）。据考证，逄国初在临淄一带。商末周初，薄姑占其地，逄国西迁至济阳一带。在济阳区曲堤街道刘台村西一高台地上，发现了西周早期的逄国贵族墓地，目前已清理4座墓葬，葬具皆为一棺一椁，出土随葬品2000余件，其中铜器32件，且有12件带有铭文，其中有9件带有“夆”字。“夆”字为国名，即“逄”。

Pang State: Liutaizi Noble Cemetery

According to legend, descendants of Emperor Yan had a tomb named Mingling, which was conferred the title of Pang in the early Shang Dynasty. According to research, Pang State was initially located in the Linzi area. At the end of the Shang Dynasty and the beginning of the Zhou Dynasty, Bo Gu occupied its territory and migrated westward to the Jiyang area. On a high platform in the west of Liutai Village, Qudi Street, Jiyang District, a noble cemetery from the early Western Zhou Dynasty was discovered. Currently, four tombs have been cleared, and the burial objects are all one coffin and one coffin. More than 2,000 burial objects have been unearthed, including 32 bronze objects, and 12 of them have inscriptions, of which 9 have the character “ 夆 ”. The character “夆”is the name of the state, which is“逄”.

国君夫人墓

刘台子6号墓是目前黄河中下游同期中型墓葬中最大的一座，出土文物1907件，其中24件铜器及916件玉器等随葬品表明墓主身份不同寻常，6鼎5簋的青铜礼器组合符合诸侯墓级别，墓主应为诸侯国国君的夫人。从一件圆鼎（M6:23）腹内壁的铭文看，逄国通过联姻，与周王室保持着密切关系。此外，该墓出土的玉器种类繁多，做工精美且玉质各异。

The Tomb of the Monarch's Wife

Liutaizi Tomb M6 is the largest mid-sized burial of its period discovered in the middle-lower Yellow River basin1. It yielded 1,907 artifacts, including 24 bronze ritual vessels inscribed with clan symbols, 916 jade ornaments, a 6-ding and 5-gui bronze set, and a ritual configuration exclusive to Zhou-era feudal lords. From the inscription on the inner wall of a round tripod (M6:23), it can be judged that Pang State maintained close relations with the Zhou royal family through marriage alliances. Furthermore, there are various types of jade artifacts unearthed from the tomb, with exquisite workmanship and different qualities of jade.

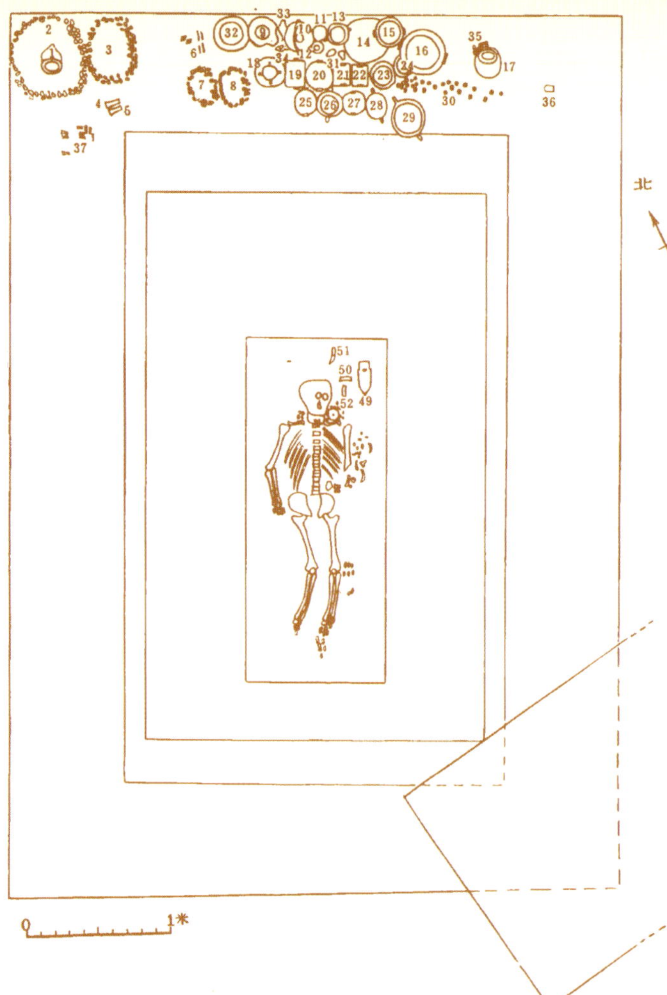

M6 平面图
Floor Plan of M6

1、32. 陶鬲
2、3、7、8. 贝圈
4、5. 骨管
6、37. 蚌饰
9、11. 铜觯
10. 铜卣
12. 铜尊
13. 铜盉
14. 铜盘
15、25、26、28、29. 铜簋
16. 铜甗
17、31. 陶罐
18. 瓷壶
19、21、22. 铜方鼎
20、23、27. 铜圆鼎
24. 铜鬲
30、35. 海贝
33、34. 铜爵
36. 卜骨
49、51. 玉戈
50. 玉鹦鹉
52. 玉璜

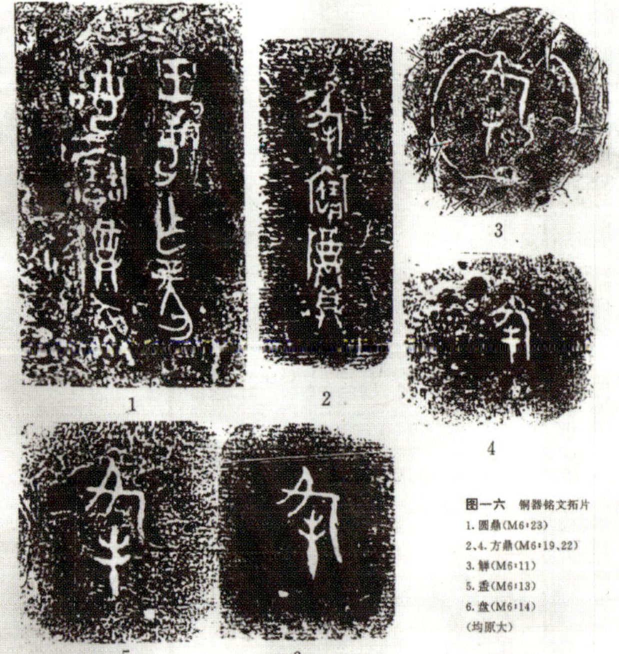

M6 出土铜器铭文拓片
Rubbings of Copperware Inscriptions Unearthed in M6

铭文：1. 王姜乍龏（gōng）姒宝尊彝　2. 夆宝尊鼎
3. 夆　4. 夆　5. 夆　6. 夆

铜方鼎

Bronze Square Tripod

西周（前 1046—前 771）
刘台子逄国贵族墓地出土
山东省文物考古研究院藏
通高 19.5 厘米，通长 18.5 厘米，通宽 14 厘米，腹深 10 厘米

器呈长方斗形。腹部中间为长方形空白，周围布满纹饰，两侧为变形夔纹，上为一周蛇纹，下为一周凤鸟纹。内壁铸有 4 字铭文“夆宝尊鼎”。

俏色玉鹅

Jade Goose

西周（前 1046—前 771）
济阳区刘台子遗址出土
济阳博物馆藏
长 3.3 厘米，高 3.2 厘米，厚 1.0 厘米

俏色品，玉色上黄下青。

玉鸳鸯

Jade Mandarin Duck

西周（前 1046—前 771）
济阳区刘台子遗址出土
济阳博物馆藏
长 3.8 厘米，宽 2.6 厘米，厚 0.2 厘米

青白色，有黄斑。全身线刻，线条流畅。

玉兔

Jade Rabbit

西周（前 1046—前 771）
济阳区刘台子遗址出土
济阳博物馆藏
长 4.2 厘米，宽 2.1 厘米，厚 0.3 厘米

青玉。曲身弓背，短尾稍卷。前后有二孔，两面钻。

俏色玉鱼鹰

Jade Osprey

西周（前 1046—前 771）
济阳区刘台子遗址出土
济阳博物馆藏
长 5.1 厘米，高 3.6 厘米，厚 0.6 厘米

俏色品。鱼鹰青黄色，质地光洁细腻，体态刚劲。

玉鹦鹉

Jade Parrot

商代晚期
济阳区刘台子遗址出土
济阳博物馆藏
长 10.8 厘米，宽 3.6 厘米

青玉，有褐斑。系鹦鹉半成品，只加工完成基本形状，未刻划细部纹饰。

青玉鱼

Sapphire Fish

西周（前 1046—前 771）
济阳区刘台子遗址出土
济阳博物馆藏
通长 8.4 厘米，通宽 2.2 厘米

墨绿色，有白斑。背上有一条长鳍，胸部和腹部各有一短鳍，鳍线较粗。

小邦邿国——仙人台邿国贵族墓地

邿（shī）国是周代东方的一个附庸小国，文献记载极少。在长清区五峰山街道北黄崖村南，出土了6座西周末期至春秋中晚期的邿国贵族墓。这一发现补充了相关的文献记载，为邿国历史的研究工作提供了珍贵的材料。墓葬葬具多为一棺一椁，棺椁均经髹漆，棺底铺撒朱砂，墓底设腰坑，内殉一狗，显示出这是一处具有共同丧葬习俗的墓地，延续时间长达200余年。6座墓葬共随葬青铜器、玉器和陶器等遗物320多件（套），其中青铜礼乐器108件（7件铸有铭文，4件铭文中有“邿”字）。规格最高的M6可能为邿国国君之墓。仙人台遗址的考古发现入选1995年度全国十大考古新发现。

The Humble State of Shi: Noble Cemetery of the State of at Xianrentai

The State of Shi was a vassal state in the eastern region of the Zhou Dynasty. There are very few records of it in literature. In the south of Beihuangya Village, Wufengshan Street, Changqing District, six noble tombs from the Western Zhou Dynasty to the mid to late Spring and Autumn Period of the State of Shi were unearthed. This discovery supplements the shortcomings of literature records and provides valuable materials for the study of the history of the State of Shi. The burial equipment in the tomb predominantly consist of an lacquered inner coffins nested within an outer coffin, with cinnabar strewn beneath the coffin bases and sacrificial canine remains interred in waist pits at the tomb floor, collectively attesting to a culturally cohesive necropolis adhering to sustained funerary conventions over 200 years. More than 320 pieces (sets) of bronze, jade, and pottery relics were buried in the 6 tombs, including 108 bronze ritual musical instruments and 7 with inscriptions, of which 4 had the character “ 邿 ” on them. The M6 with the highest specification may be the tomb of the monarch of the State of Shi. The archaeological discoveries of Xianrentai Site had been selected as one of the top ten new archaeological discoveries in China in 1995.

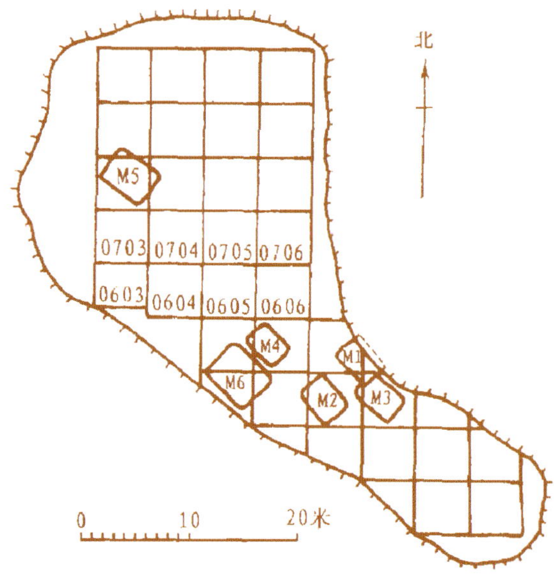

仙人台周代墓地墓葬分布图
Distribution Diagram of Tombs of the Zhou Dynasty at Xianrentai

青铜方壶

Bronze Square Pot

春秋（前 770—前 476）
仙人台邿国贵族墓地出土
山东大学博物馆藏
口径 32 厘米，高 19.5 厘米

酒器。通体扁方，口微敞，长径稍内束，鼓腹下垂，圈足。饰垂鳞纹、浅浮雕龙纹、环带纹、宽带纹、蟠龙纹，颈两侧有凤鸟衔环状耳，环呈索状。工艺复杂，特别是衔环处，应为凤首和衔环分别铸造，再与器身合铸而成。方壶的铸造工艺代表了古代铸铜工艺的最高水平。

邿公典盘

Ritual Basin (Dian Pan) of the Duke of Shi

春秋（前 770—前 476）
仙人台邿国贵族墓地出土
山东大学博物馆藏
口径 43.2 厘米，通高 7.2 厘米

水器。盘内底有铭文 6 行 42 字，重文 3 字。出土于仙人台 M5。M5 墓主人为一成年女性。从其葬具为一棺一椁，以及三鼎、二敦（duì）和编钟、编磬等成套礼乐器及两车等随葬器物来看，墓主应是士一级的贵妇人。从邿公典盘铭文分析，墓主应是某一姜姓国女嫁给邿国王室为妻者。

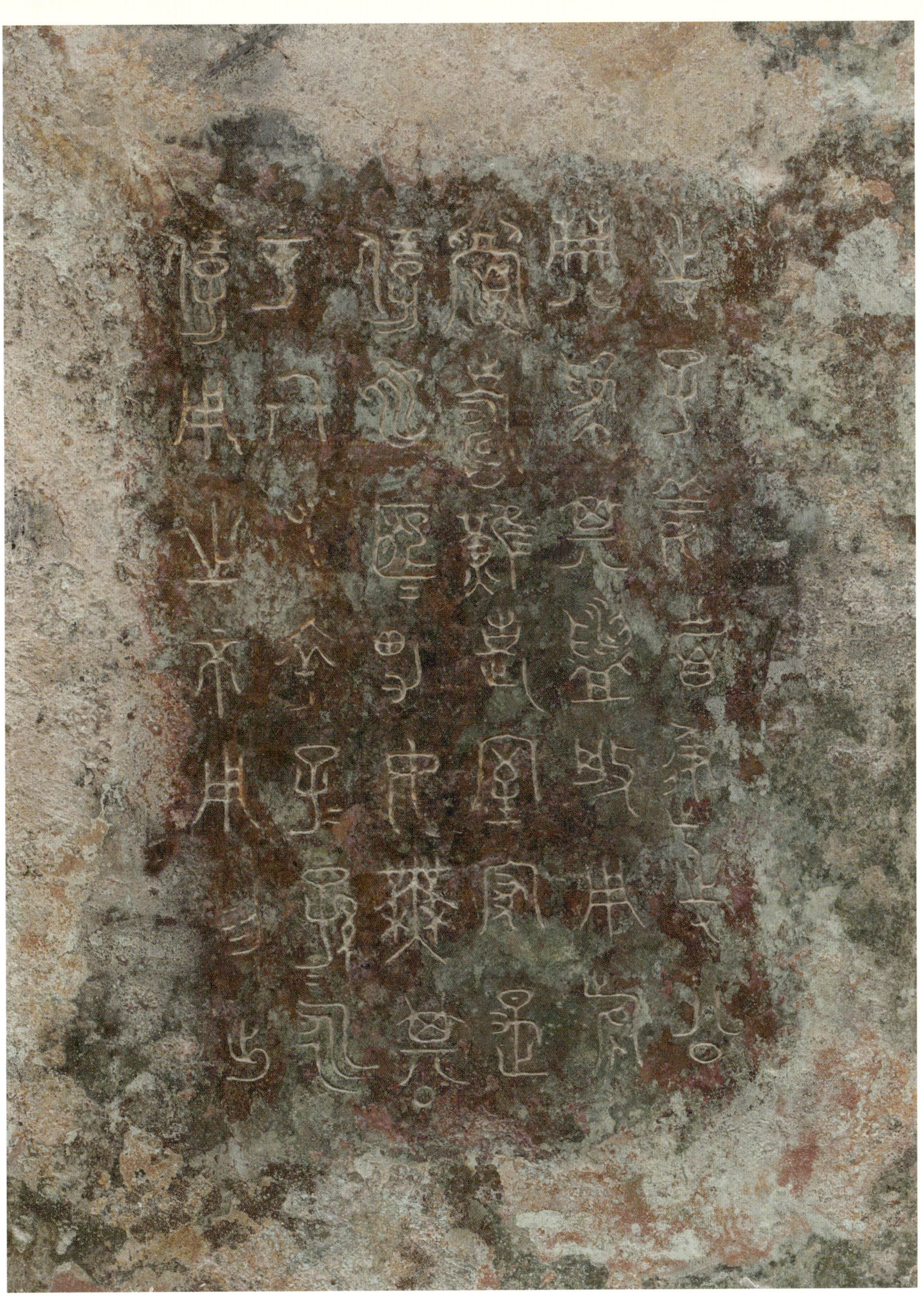

青铜簠（fǔ）

The Bronze Fu

春秋（前 770—前 476）
仙人台邿国贵族墓地出土
山东大学博物馆藏
口径 32 厘米，宽 26 厘米，高 19.5 厘米

盛食器。共两件，大小、形制基本相同。器盖、器身造型一致，方唇，长方形大敞口，斜直腹，腹侧各有一兽耳钮，平底，圈足外卷，器表饰象首纹。器内顶部和底部分别有铭文，为对文。铭文为“邿召乍为其旅簠，用实旅梁，用飤诸母诸兄，使爰宝母又疆”。

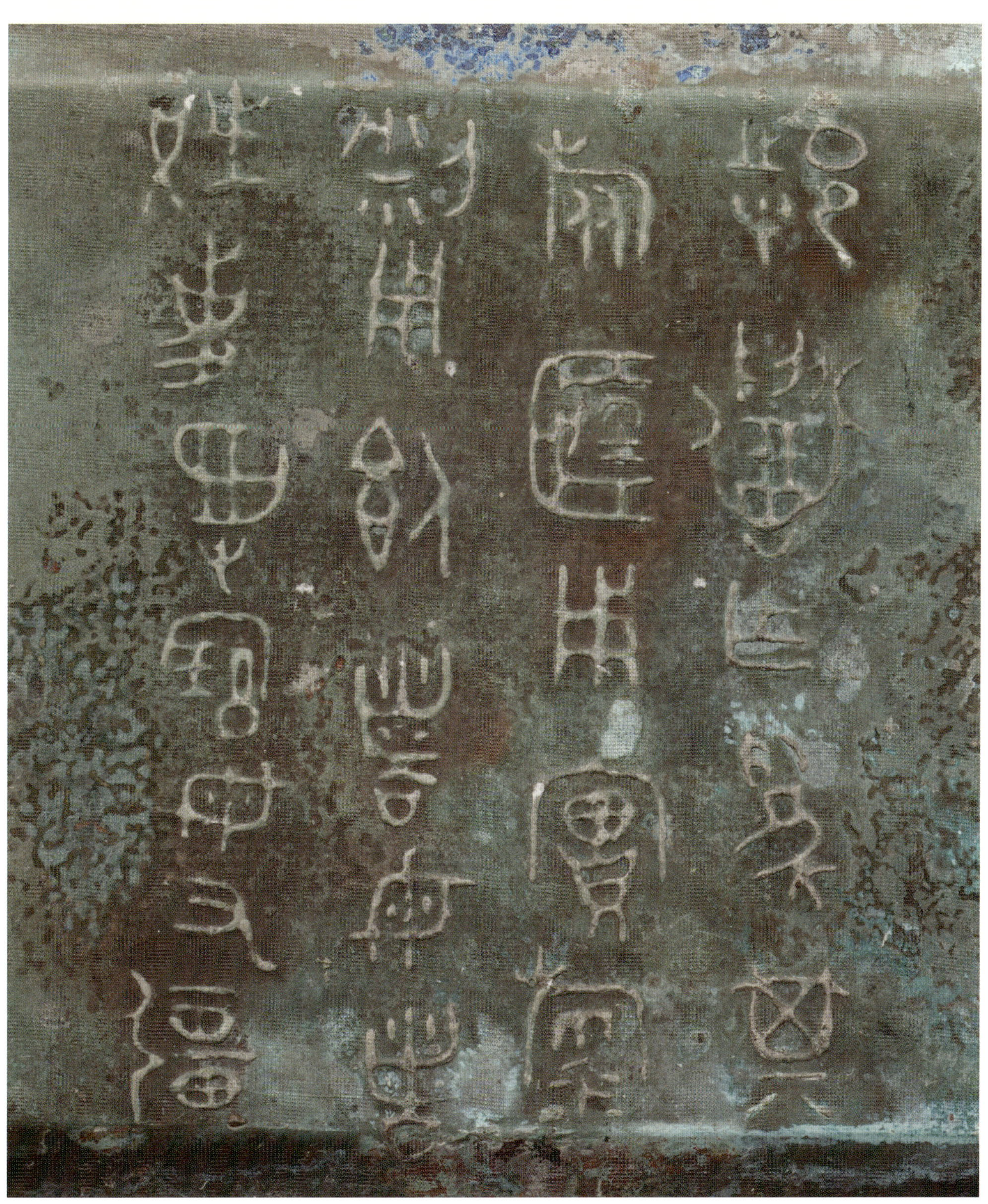

齐风浩荡

章丘女郎山战国墓

章丘区绣惠街道北部的女郎山墓地出土了一座战国中期的齐国大墓。该墓上部原有高达10余米的封土，全墓平面呈“甲”字形，葬具为一棺两椁，棺、椁表面均髹有红、黑、黄漆。主墓室随葬品极为丰富，包括29件青铜礼器、105件车马器和10余件铜乐器。根据墓葬形制、随葬品组合以及当时作为齐国都城临淄的管辖地域等特征判断，该墓主极有可能是《战国策·齐策六》中记载的齐威王、宣王时期的重要军事将领匡章。

The Majestic Grandeur of Qi ethos

Warring States Period Tomb Complex at the Nvlang Mountain Cemetery in Zhangqiu

A large tomb of Qi State from the mid Warring States period has been unearthed at the Nvlang Mountain Cemetery in the north of Xiuhui Street, Zhangqiu District, Jinan City. The upper part of the tomb was originally covered with soil up to more than 10m in height, and the entire tomb plane is in the shape of Chinese character “甲”. Thc burial chamber is a single inner coffin enclosed by two outer coffins, with all wooden surfaces coated in alternating polychrome lacquer layers of vermilion, jet black, and ocher pigments. The main tomb is extremely rich in burial objects, including 29 bronze ritual vessels, 105 chariot and harness fittings, and over 10 bronze musical instruments. Based on the structure of tombs, the combination of burial objects, and the jurisdictional area of Linzi, which was the capital city of Qi at that time, it is highly likely that the tomb owner was Kuang Zhang, an important military general during the reigns of King Wei and King Xuan of Qi, as recorded in the *Strategies of the Warring States: Qi Strategy VI*.

乐舞陶俑线稿
Line Chart of Music and Dance Terracotta Warriors

匡章

匡章初为齐威王将领，曾率军击退秦军进攻。齐宣王六年（前 314），齐国乘燕国内乱，派匡章率兵十万攻燕，50 日内直破燕都。齐湣王时，齐国联合韩、魏攻楚，匡章率军大败楚军。齐湣王三年（前 298），匡章率齐、魏、韩联军攻破函谷关，迫使秦国求和。

Kuang Zhang

Kuang Zhang was originally a general of King Qi Wei and led his army to repel the Qin army's attack. In the sixth year of King Xuan of Qi (314 BC), the State of Qi took advantage of the chaos within Yan and sent Kuang Zhang to lead 100,000 troops to attack Yan. Within fifty days, they broke through the capital of Yan. During the reign of King Min of Qi, the State of Qi allied with the State of Han and the State of Wei to attack the State of Chu, and Kuang Zhang led his army to a great victory over the Chu army. In the third year of King Min of Qi (298 BC), Kuang Zhang led the coalition forces of the State of Qi, the State of Wei and the State of Han to break through the Hangu Pass and force the State of Qin to make peace.

彩绘乐舞陶俑

Polychrome-painted Music and Dance Terracotta Figurines

战国（前 476—前 221）
女郎山战国墓出土
山东省文物考古研究院藏

这一组彩绘乐舞陶俑共38件，其中人物俑26件、乐器4种、祥鸟8只。人物俑面施粉红彩，衣纹褶皱清晰，造型生动，动作刻划逼真细腻，被誉为“神形兼备、生活气息最为浓郁的战国陶塑佳作”，为研究齐国乐舞、服饰提供了珍贵的实物资料。

长清岗辛战国墓

长清岗辛战国墓是一座战国晚期齐墓，位于长清区归德街道岗辛村北。发掘时还保留着高6.8米、底径约60米的椭圆形封土。墓的形制与齐都临淄战国大型墓一致。椁室用天然石块垒砌而成，以卵石充填缝隙。椁室东南隅二层台上有一器物库，出土铜鼎、豆、壶、舟、弩机、帷架构件和铜明器鼎、钫、盘、匜、罐、食盒等近百件随葬品，其中以错铜丝镶绿松石铜豆和帷架构件最为珍贵。

Gangxin Tomb of the Warring States Period at Changqing

Gangxin Tomb of the Warring States Period at Changqing is a Qi tomb from the late Warring States period, located north of Gangxin Village, Guide Street, Changqing District. There is an elliptical soil seal on the ground with a height of 6.8m and a bottom diameter of approximately 60m. The shape and structure of the bomb are consistent with those of the large-scale tombs of the Warring States Period in Linzi, the capital of Qi. The outer coffin chamber was constructed with natural stone blocks in dry-stone masonry, with interstices systematically packed with fluvial pebbles. A funerary object repository is structurally integrated into the second-tier ledge within the southeastern quadrant of the outer coffin chamber. The burial assemblage comprises nearly a hundred ritual bronzes and funerary objects, including bronze ding tripods, dou stemmed dishes, hu jars, and zhou wine vessels, crossbow triggers, canopy frame elements with curtain hooks, bronze fang square vessels, pan basins, yi ewers, jars and food containers. Among the burial objects, the dou stemmed dishes with bronze wire decoration and the bronze canopy frame components are of particular archaeological significance.

错铜丝镶绿松石豆

Turquoise-inlaid Bronze Dou (Ceremonial Stemmed Vessel) with Bronze Wire Inlay

战国（前 476—前 221）
岗辛战国大墓出土
山东省文物考古研究院藏
通高 27 厘米，口径 18.3 厘米，圈足径 13.3 厘米

豆盘呈半球状，子口，深圆腹，柄略高，喇叭形圈足。覆钵形盖，扁平握手，有短柄，盖面及器身均饰由黄铜丝与绿松石镶嵌而成的几何形勾连云纹。

梁二村战国墓

梁二村战国墓的发现为研究战国时期齐国葬俗、齐国边邑、齐文化变迁、齐国历史等提供了重要资料。梁二村战国墓位于历城区鲍山街道北，发现墓葬 3 座，其中两座大墓东西相距 150 米，均为竖穴土坑积石木椁墓，时代大致为战国晚期。M1 为“甲”字形墓，出土器物 40 余件（组），相对完整的有青铜编钟 19 件、句鑃 9 件等。根据形制和出土器物推测，M1、M3 墓主当为大夫一级的贵族；M2 墓主为士级别，应为陪葬墓或家族墓。

Warring States Tomb in Liang'er Village

The discovery of the tomb of Qi nobles in Liang'er Village provides important information for studying the burial customs, border towns, cultural changes, and history of Qi during the Warring States period. The Liang'er Village Warring States Tombs are located in the north of Baoshan Street, Licheng District. There are three tombs, two larger tombs are 150m apart from east to west, both of which are vertical-cave earth pit stone and wooden coffin tombs, dating back to the late Warring States period. M1 is a tomb in the shape of Chinese character “ 甲 ”, with over 40 unearthed artifacts (sets), including 19 relatively complete bronze bronze chimes and 9 bronze pendants. Based on the shape and unearthed artifacts, it is speculated that the owner of M1 and M3 tombs should be noblemen of the doctor level; the owner of the M2 tomb should be a nobleman level, and it should be a funerary tomb or a family tomb.

春秋时期木框架水井
Wooden Framed Well of Spring and Autumn Period

济南梁二村战国墓 M1 发掘现场
Excavation Site of M1 Noble Tomb of Qi State in Liang’er Village

变形龙纹铜镈（bó）

Deformed Large bronze Bell with Dragon Patterns

战国（前476—前221）
梁二村战国墓出土
历城区博物馆藏

编钟是中国古代的一种大型打击乐器，变形龙纹铜镈表面装饰变形的龙纹，龙纹形态抽象且富有动感，展现了古代工匠的高超技艺。铜镈整体造型庄重，纹饰精美，具有较高的艺术和历史价值，是研究古代青铜文化和纹饰演变的重要实物资料。

蟠（pán）螭（chī）纹铜钮钟

Bronze Bo Bells with Stylized Dragonpatterns

战国（前 476—前 221）
梁二村战国墓出土
历城区博物馆藏

蟠螭纹铜钮钟属于编钟的一种，通常为合瓦形钟体，底部弯曲，顶部为桥形口，钟体呈椭圆形或合瓦形。钟体上装饰精美的蟠螭纹，纹饰细腻且规整，表现出春秋晚期青铜器纹饰的高超工艺水平。

透雕变形龙凤纹铜方镜

Openwork Square Bronze Mirror with Stylized Dragon-Phoenixpatterns

战国（前476—前221）
梁二村战国墓出土
历城区博物馆藏
通长13.4厘米，通宽13.4厘米，厚7厘米，重350克

透雕变形龙凤纹铜方镜采用透雕工艺，装饰变形的龙凤纹。龙纹象征权威，凤纹象征吉祥，变形设计则体现了艺术创新。该文物具有较高的历史和艺术价值，反映了当时的工艺水平和文化内涵。

瓦楞纹铜盖豆

Bronze Covered Dou Vessel with Corrugated Fluting

战国（前 476—前 221）
梁二村战国墓出土
历城区博物馆藏
通高 46.5 厘米，腹径 19.7 厘米

瓦楞纹铜盖豆由盖和豆两部分组成，整体呈圆弧形，盖与豆通过子母口相扣合。作为古代盛食器，主要用于祭祀或宴饮活动。其设计注重实用性与美观性相结合，体现了战国时期青铜器工艺的高超水平。

齐鲁联姻

春秋时期，联姻是各国巩固关系的重要手段。鲁国与齐国之间的联姻历史悠久且颇为频繁。据记载，从鲁桓公至鲁成公连续7任鲁公中，有6位的正夫人均娶于齐国，这种联姻关系在两国之间的合作与对抗中发挥了重要作用。

The Intermarriage between the States of Lu and Qi

During the Spring and Autumn period, intermarriage was an important means for countries to consolidate their relationships. The intermarriage between the states of Lu and Qi has a long and frequent history. According to records, from Duke Huan of Lu to Duke Cheng of Lu, of all the seven consecutive Dukes of Lu, six legitimate wives were married to the State of Qi. This marriage alliance played an important role in the cooperation and confrontation between the two countries.

齐叔姬双耳铜盘

Qi Shuji's Bronze Basin with Bilateral Cast-On Handles

春秋（前770—前476）
济南市博物馆藏
口径46.4厘米，底径36厘米，高12.7厘米

盥器。敞口，双附耳，浅腹，平底，矮圈足。腹饰蟠螭纹，间以涡纹，耳饰重环纹，足为垂鳞纹。盘内底有铭文“齐叔姬作孟庚宝盘，其万年无疆，子子孙孙永受大福用”4行22字。“叔姬”，乃鲁国国君之女，嫁为齐昭公夫人。此盘为齐国铸器，其制作年代当是昭公在位之时。

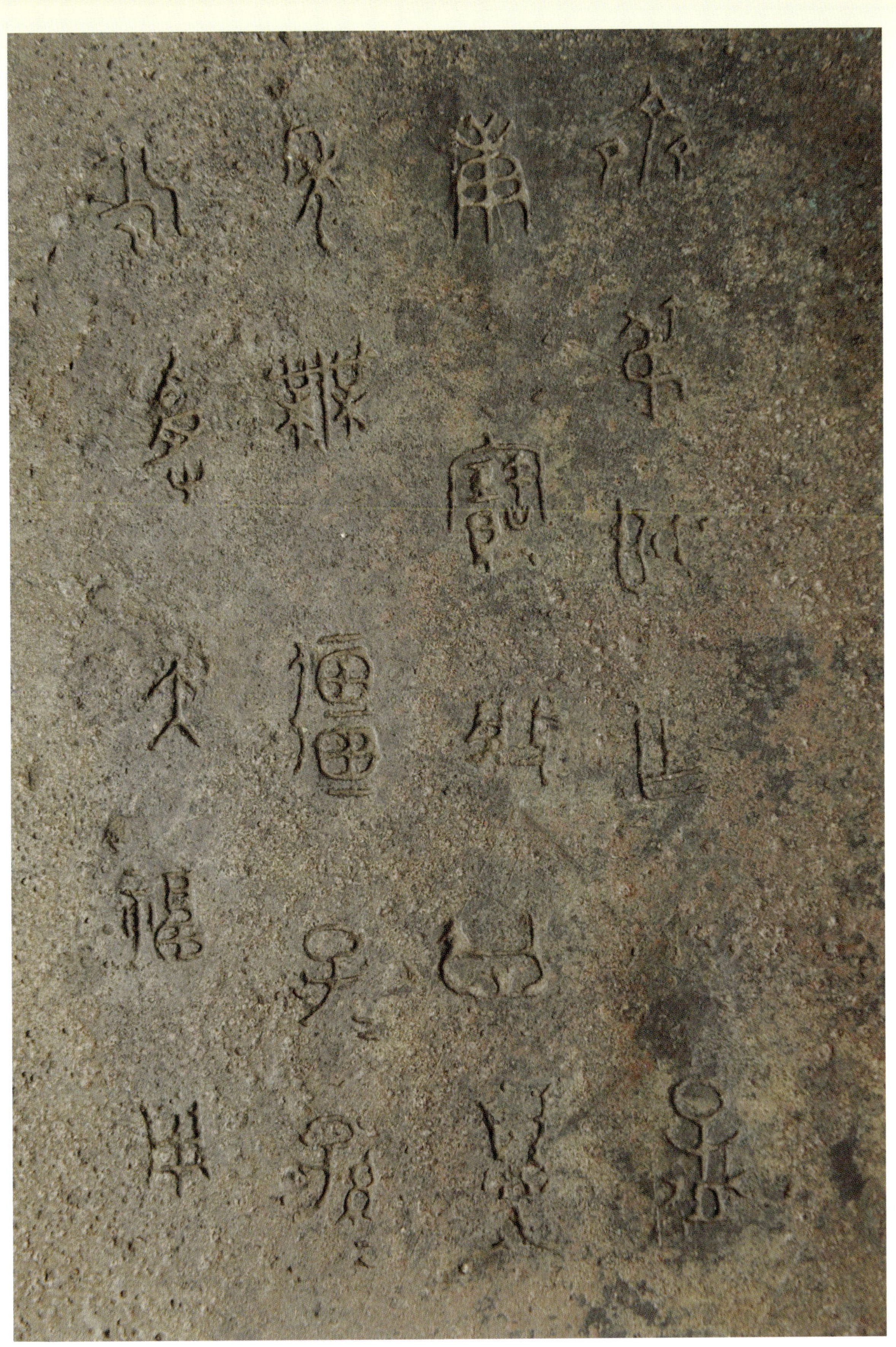

郡国并行（汉代）

秦初并天下时，济南地区大部属济北郡。汉初，分济北郡置博阳郡，后博阳郡治迁至济水之南的东平陵，博阳郡改称济南郡，“济南”一名始见诸史册。汉代实行郡国并行制，汉文帝十六年（前164），设济南国。汉景帝三年（前154），废济南国，复置济南郡。汉武帝时，济南郡辖东平陵、历城等14县，属青州刺史部。东汉建武十七年（41），济南郡复称济南国，辖14县，后改辖10县。汉代济南地区的农业和工商业十分发达，东平陵是一座闻名遐迩的繁荣都市。

The Parallel System of Prefectures and States

During the Qin Dynasty's initial territorial consolidation, the majority of the Jinan region was administratively incorporated into Jibei Prefecture. At the beginning of the Han Dynasty, Jibei Prefecture was divided and Boyang Prefecture was established. Later, county town of Boyang Prefecture was relocated to Dongpingling, south of the Jishui River. Boyang Prefecture was renamed Jinan Prefecture, and the name "Jinan" first appeared in annals of history. The Han Dynasty implemented a parallel system of prefectures and states, and in the 16th year of Emperor Wen of Han (164 BC), the State of Jinan was established. In the third year of Emperor Jing of Han (154 BC), the State of Jinan was abolished and Jinan Prefecture was reestablished. During the reign of Emperor Wudi of the Han Dynasty, Jinan Prefecture administered 14 counties including Dongpingling and Licheng, under the jurisdiction of the Qingzhou Department of State. In the 17th year of the Jianwu reign of the Eastern Han Dynasty (41 AD), Jinan Prefecture was renamed the State of Jinan, with 14 counties under its jurisdiction, and later changed to 10 counties. During the Han Dynasty, agriculture and commerce were highly developed in the Jinan area, and Dongpingling was a renowned and prosperous city.

“济北守印”封泥

Clay impression of Seal of “Official Seal of Prefectural Administrator of Jibei”

西汉（前 202—8）
济南市博物馆藏

封泥，是古代抑印于泥用以封缄的遗存，主要用于封简牍文书，也用于封坛罐、囊箱等，能够起到封护、保密、防止非法启封的作用。

济北，西汉时期郡国名。“守”为地方行政长官“郡守”之简称。此“济北守印”封泥，称“守”而不称“太守”，应是汉景帝中元二年（前 148）以前之物。

吕国——洛庄汉墓

汉初，吕后以其兄之子吕台为吕王，建都平陵城。位于东平陵故城以东6公里的洛庄汉墓，据推测可能是吕国第一代王吕台之墓。遗址范围东西约200米、南北100余米，地上原有巨大方形封土。洛庄汉墓是一座有东西墓道的“中”字形竖穴木椁墓，环绕墓室发现陪葬坑和祭祀坑共36座，陪葬坑中出土数以千计的文物。洛庄汉墓的考古发现入选2000年度全国十大考古新发现。

The State of Lv: The Han Tomb at Luozhuang

In the early Han Dynasty, Queen Lv took his brother's son Lv Tai as King Lv and established the capital city of Pingling. The Han Tomb at Luozhuang, located 6km east of the old city of Dongpingling, is speculated to be the tomb of Lv Tai, first king of the State of Lv. The site covers an area of about 200m from east to west and over 100m from north to south, with a huge square-shaped mound on the ground. The Han Tomb at Luozhuang is a Chinese character “中”shaped vertical cave wooden coffin tomb with east-west tomb passages. A total of 36 funerary pits and sacrificial pits were found around the tomb, and thousands of cultural relics were unearthed in the funerary pits. The archaeological discoveries of The Han Tomb at Luozhuang had been selected as one of the top ten new archaeological discoveries in China in 2000.

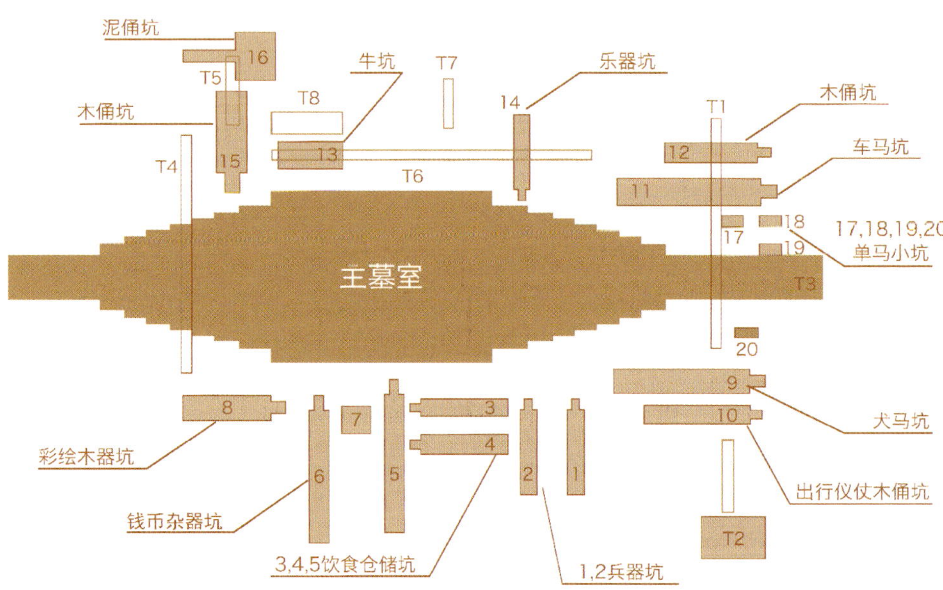

洛庄汉墓平面布局示意图
Schematic Diagram of the Plane Layout of The Han Tomb at Luozhuang

14号坑编钟、编磬出土状况
Unearthed Condition of Bronze Chimes and Stone Chimes of Pit 14

洛庄汉墓乐器

洛庄汉墓14号陪葬坑出土了保存完好的编钟一套19件、石磬6组107件等140余件乐器，堪称“汉代乐器大观”，是迄今为止集中出土古代乐器最多的一次。洛庄汉墓乐器坑不仅是考古发现的第一个专门的乐器陪葬坑，而且坑内出土的编磬数量超过了国内历次汉代考古发现的实用编磬数量的总和，这为研究汉初的礼乐制度和乐器组合提供了新的资料。

Unearthed Musical Instruments of The Han Tomb at Luozhuang

More than 140 musical instruments, including a well-preserved set of 19 bronze chimes and 6 sets of 107 stone chimes, were unearthed from the No.14 burial pit of the Han Tomb at Luozhuang. It can be regarded as a grand view of musical instruments of the Han Dynasty and is the largest collection of ancient musical instruments unearthed so far. The Musical Instrument Pit of the Han Tomb at Luozhuang is not only the first dedicated burial pit for musical instruments discovered in archaeology, but also the number of unearthed stone chimes is more than the total number of practical stone chimes discovered in previous Han Dynasty archaeological discoveries in China, which provides new information for studying the ritual and music system and musical instrument combinations in the early Han Dynasty.

铜钲（zhēng）

Bronze Bell-shaped Percussion Instrument (Used in Ancient Times by Troops on March)

西汉（前 202—8）
洛庄汉墓陪葬坑和祭祀坑遗址出土
章丘区博物馆藏
长 16.5 厘米，宽 12.4 厘米，高 30.5 厘米

古代乐器，又名“丁宁”，形似钟而狭长，有长柄可执，击之而鸣，是行军乐器。此铜钲为合瓦形，甬为空心圆筒状，一侧有一小圆孔用以引绳，钲两面为素面，器壁较薄。

青铜铃铛

Bronze Bell

西汉（前 202—8）
洛庄汉墓陪葬坑和祭祀坑遗址出土
章丘区博物馆藏
直径 3 厘米，高 3 厘米

出土于瑟钥旁，发现时其上有红绳痕迹，原应为一组。铃铛大小相同，均为圆球形，上面镂孔，内置金属核，摇之作响。

石编磬

Stone Chime

西汉（前 202—8）
洛庄汉墓陪葬坑和祭祀坑遗址出土
章丘区博物馆藏
最大：长 62 厘米，宽 25 厘米，厚 2.3 厘米
最小：长 21.9 厘米，宽 10.6 厘米，厚 2.2 厘米

“磬，乐器也；以玉或石为之，其形如矩。”磬，古代打击乐器之一，多用于宫廷雅乐或盛大祭典。编磬均为青石质地，呈曲尺形，刻有“鲁加”“益瓦”等铭文字样，为悬挂顺序及定音、测音提供了直接依据。

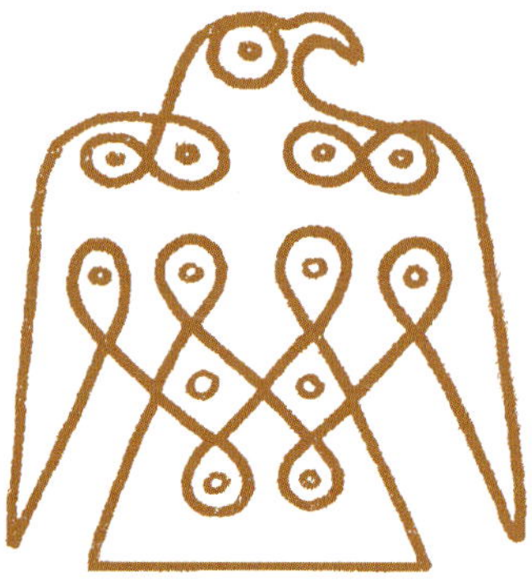

铜錞（chún）于

Bronze Chunyu（Percussion Instruments Used for Military Command in Ancient China）

西汉（前 202—8）
洛庄汉墓陪葬坑和祭祀坑遗址出土
济南市考古研究院藏
长 25.8 厘米，宽 22.6 厘米，高 48.3 厘米

古代军中所用铜制打击乐器。环钮，圆肩，束腰，椭圆形平口微敞。腰部两侧分别浮雕一鹰，鹰的轮廓为一笔画成，甚是罕见。

洛庄汉墓车马器

洛庄汉墓 11 号坑为车马坑，共埋葬 3 辆大车，每车驷马，车马饰件齐全，马具多为鎏金制品，尽展墓主生前之奢华。许多车马具都有浓厚的秦代风格，有的甚至与秦始皇陵铜车马的装具相同，因此，这一规模宏大的车马坑被称为“第二秦陵”，对于研究汉初的车马制度具有重要价值。

Unearthed Carriages and Horses of The Han Tomb at Luozhuang

Pit 11 of the Han Tomb at Luozhuang is a carriage and horse pit, with a total of three large carriages buried, each with four horses. The carriages and horses are fully decorated, and the horse gear is mostly gilded, showing the luxury of the tomb owner during his lifetime. Many carriages and horses have a strong style of the Qin Dynasty, some of which are even the same as those of the bronze chariots and horses in the Mausoleum of the First Qin Emperor. Therefore, this large-scale chariot and horse pit is called the “The Second Mausoleum of the Qin Emperor”, which is of great value for the study of the chariot and horse system in the early Han Dynasty.

车马结构部件示意图
Schematic Diagram of Chariot and Horse Structural Components

11号车马坑发掘现场
Excavation of Shafts, Yokes and Horse Skeletal Remains in Pit 11

鎏金当卢

Gilt Damloup

西汉（前 202—8）
洛庄汉墓陪葬坑和祭祀坑遗址出土
济南市考古研究院藏
长 16.5 厘米，宽 7.8 厘米，厚 1.4 厘米

当卢，是古代系于马头部的饰件，放置于马的额头中央偏上部。此西汉当卢，其形如叶，镂空浮雕，主题图案为一匹卷曲呈反“S”形的骏马，辅以变化的鸟纹和云纹，造型优美，实属罕见艺术佳作。

金带扣
Gold Buckle

西汉（前 202—8）
洛庄汉墓陪葬坑和祭祀坑遗址出土
济南市考古研究院藏

带扣，是一种用于连接和固定车马具的配件，兼具装饰和稳固功能。

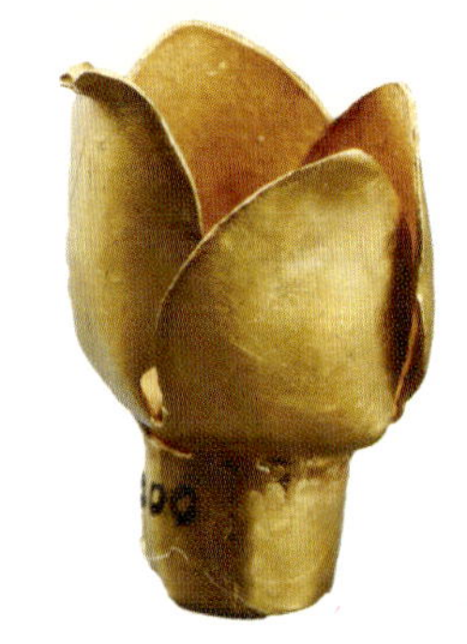

金栓
Peg of Metal

西汉（前 202—8）
洛庄汉墓陪葬坑和祭祀坑遗址出土
济南市考古研究院藏

金栓，推测为一种装饰车马部件的小塞子。

金环
Gold Ring

西汉（前 202—8）
洛庄汉墓陪葬坑和祭祀坑遗址出土
济南市考古研究院藏

车马器，金质实心。推测为连接或装饰车马配饰之用。

金节约
Peg of Metal

西汉（前 202—8）
洛庄汉墓陪葬坑和祭祀坑遗址出土
济南市考古研究院藏

用于连接马具的各个部分，起到节制和约束马匹的目的，因此得名“节约”。

金泡
Gold Bubble

西汉（前 202—8）
洛庄汉墓陪葬坑和祭祀坑遗址出土
济南市考古研究院藏

金泡，是一种装饰在马头上的饰品，形态多样。其背后设有钮，便于穿入皮革之中。

鹿首形金节约
Peg of Metal of Deer Head-shaped

西汉（前 202—8）
洛庄汉墓陪葬坑和祭祀坑遗址出土
济南市博物馆藏

由金片锤揲而成，鹿角卷曲勾环相连，鹿耳灵动，鹿目圆睁。

青铜龙马纹马镳（biāo）

Bronze Harness Component with Dragon and Horse Patterns

西汉（前202—8）
洛庄汉墓陪葬坑和祭祀坑遗址出土
章丘区博物馆藏
长12.2厘米，宽4.4厘米，厚0.7厘米

镳，与衔配套使用，贯于衔两端，显露于马口两旁。

错金银铁马具（桃形）

Iron Horse Harness with Gold and Silver Inlay (In Peach Shape)

西汉（前 202—8）
洛庄汉墓陪葬坑和祭祀坑遗址出土
济南市考古研究院藏
长 12.8 厘米，宽 9.4 厘米

两桃形上下叠压，上下均饰对称双兽图案，呈腾跃弹跳状。

错金银铁马具（圆形）

Iron Horse Harness with Gold and Silver Inlay (in Round Shape)

西汉（前 202—8）
洛庄汉墓陪葬坑和祭祀坑遗址出土
济南市考古研究院藏
直径 10.8 厘米，厚 0.6 厘米

内饰首尾相继的双兽，犄角高昂，长尾飘飘，奔腾如飞。

错金银铁马衔

Horse-bit with Gold and Silver Inlay

西汉（前 202—8）
洛庄汉墓陪葬坑和祭祀坑遗址出土
济南市考古研究院藏
长 23.5 厘米，单个宽 2.5 厘米，拉直时宽 20 厘米，中间环直径 4.0 厘米

马衔使用时置于马口内，外部两环连接缰绳，起到控制方向的作用，是人类驯化、驾驭马匹的重要标志。

洛庄汉墓生活用具

洛庄汉墓除了陪葬的乐器、车马器，还出土了大量生活用具。在35号陪葬坑中，3号壶内盛有鱼骨，保存较好；7、8、9号壶上覆盖成片骨骼，多为禽类的肢骨；西部南北向分别放置羊骨和鱼骨。

Unearthed Living Utensils of The Han Tomb at Luozhuang

Except for the funerary musical instruments, chariots and horse equipment, a large number of living utensils were also unearthed of The Han Tomb at Luozhuang. In the funerary pit No.35, there are fish bones in pot No. 3, which are well-preserved; Pots No.7, No.8 and No.9 are covered with pieces of bones, mostly limb bones of birds. Sheep and fish bones are respectively placed in the north and south direction in the west.

洛庄汉墓35号陪葬坑内的生活用具
Living Utensils in the Funerary Pit 35 of he Han Tomb at Luozhuang

吕姓诸侯王墓

洛庄汉墓出土遗物具有西汉初年的特点，根据陪葬坑中出土的“吕大官印”“吕内史印”等封泥，并结合文献记载推断，墓主可能是死于公元前187年的吕国第一代王吕台。如果这一推断不误，那么洛庄汉墓是迄今发现的唯一一座吕姓诸侯王墓。

Tomb of the Feudal Prince Surnamed Lv

The Luozhuang Han Tomb exhibits funerary artifacts characteristic of the early Western Han Dynasty. Based on clay impression of seals inscribed with “Seal of the Grand Official of Lv” and “Seal of the Internal Secretary of Lv” discovered from funerary burial pits, coupled with historical records, scholars hypothesize the tomb owner to be Lv Tai, the first emperor of the State of Lv, who died in 187 BCE. If the inference is correct, then the Han Tomb at Luozhuang should be the only tomb of feudal princes surnamed Lv discovered so far.

“吕大官印”封泥

Clay Impression of Seal

西汉（前202—8）
章丘区洛庄汉墓出土
济南市考古研究院藏
长3.5厘米，宽3.5厘米，厚1.3厘米

方形，上有“吕大官印”篆书印文。

“吕内史印”封泥

Clay Impression of Seal

西汉（前202—8）
章丘区洛庄汉墓出土
济南市考古研究院藏
左：长3.3厘米，宽3.1厘米，厚0.9厘米
右：长3.3厘米，宽3.2厘米，厚1.5厘米

近圆形，上有“吕内史印”篆书印文。

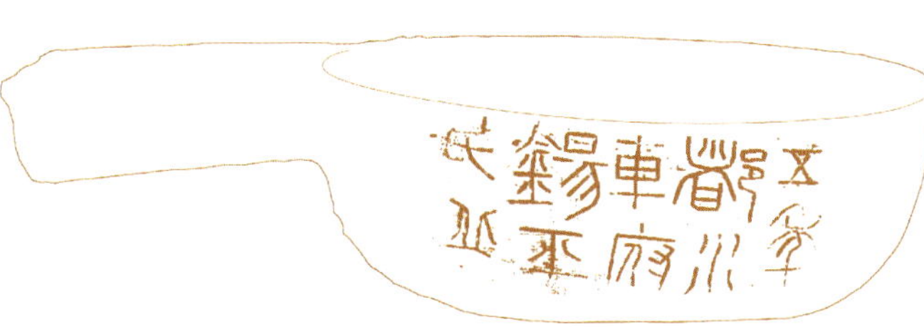

青铜量

Bronze Measuring Vessel

西汉（前 202—8）
章丘区洛庄汉墓出土
章丘区博物馆藏
长 17.7 厘米，宽 6.4 厘米，高 4.7 厘米

量，是古代社会计量农产品的主要器具。《汉书·律历志》："量者，龠、合、升、斗、斛也，所以量多少也……合龠为合，十合为升，十升为斗，十斗为斛……职在太仓，大司农掌之。"此青铜量外刻十字铭文"五秀都水车府，锡平氏丘"。

青铜鼎（带盖）

Bronze Tripod (With Lid)

西汉（前 202—8）
章丘区洛庄汉墓出土
章丘区博物馆藏
长 26.6 厘米，宽 23.8 厘米，高 23.4 厘米

弧形盖，盖面有 3 个环纽，双附耳，子母口，扁圆腹，腹部中部有一周凸棱。圜底，兽蹄状足。鼎为烹煮肉食、实牲祭祀和宴飨的食器，一般三足两耳。古代视其为立国重器、权力象征。

铁炉

Iron Furnace

西汉（前 202—8）
章丘区洛庄汉墓出土
章丘区博物馆藏
长 53.4 厘米，宽 44 厘米，高 19.6 厘米

铁炉，是汉代的一种生活用具，在各地遗址和墓葬中多有出土。此件铁炉出土于洛庄汉墓仓储类陪葬坑，主要用以烧烤食物。

鸡蛋

Egg

西汉（前 202—8）
洛庄汉墓陪葬坑和祭祀坑遗址出土
济南市考古研究院藏
长径 4.9 厘米，短径 3.8 厘米

洛庄汉墓出土鸡蛋，其中一枚完整，其余为蛋壳碎片。

青铜匜（yí）

Bronze Yi (Vessel in a Ladle Shape)

西汉（前 202—8）
洛庄汉墓陪葬坑和祭祀坑遗址出土
章丘区博物馆藏
长 37.8 厘米，宽 28.4 厘米，高 14.2 厘米

《左传》有“奉匜沃盥”之说，“沃”的意思是浇水，“盥”的意思是洗手洗脸，这说明匜是一种盥洗用具。早期匜为青铜制，汉代以后出现匜金银器、匜漆器、匜玉器。

郡治王都——东平陵城

东平陵在汉代曾为吕国、济南国的都城及济南郡的郡治所在。汉代东平陵城人口众多、手工业发达，是济南地区的政治、经济中心。东平陵故城位于章丘区龙山街道阎家村北，是山东省乃至中国保存状况最好的战国至汉代地方城址之一。故城城址呈正方形，城内面积约 400 万平方米。城址四周可见残存的夯土城墙，发现了 4 座城门遗迹。遗址中部西侧 300—400 米的范围内为冶铁遗址；南面为陶窑群，是城内手工业区；东部偏北处发现大面积的夯土基址和铺砖石的路面，当地群众称之为“殿基地”，为宫殿区；西北部及其余部分为居民区。

County Town & Capital City: Dongpingling City

Dongpingling was once the capital of the State of Lv and the State of Jinan, as well as the County Town of Jinan Prefecture during the Han Dynasty. During the Han Dynasty, Dongpingling City had a large population and developed handicrafts, making it the political and economic center of Jinan region. Old City of Dongpingling is located in the north of Yanjia Village, Longshan Street, Zhangqiu District. It is one of the best preserved local city sites from the Warring States period to the Han Dynasty in Shandong Province and even in China. The site of the old city is square, with an area of about 4 million m^2 in the city. Remnants of rammed earth city walls can be seen around the city site, and remnants of four city gates have been discovered. The iron smelting site is located within a range of 300-400m to the west of the central part of the site. In the south is the pottery kiln group, which was the handicraft area within the city. A large area of rammed earth foundations and paved roads with bricks and stones were discovered in the region to the northeast of the east. The local people called it the “palace base”, which is the palace area. Residential areas are in the northwest and other parts.

*全国重点文物保护单位
National Key Cultural Relics Protection Unit

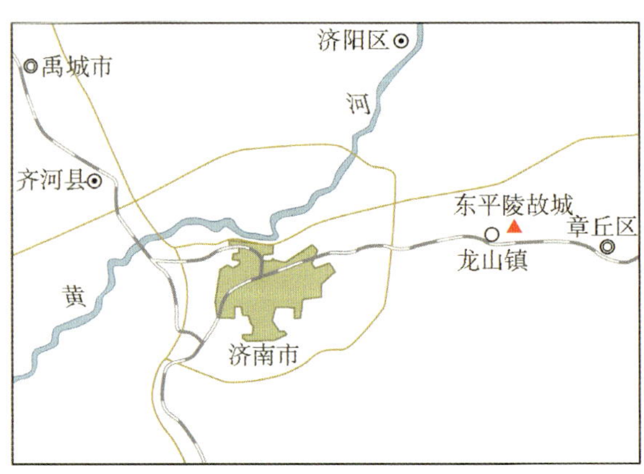

东平陵故城遗址位置示意图
Location map of the Dongpingling Mausoleum Ruins

“东平陵丞”封泥

Clay Impression of Seal

汉（前202—220）
东平陵城遗址出土
山东博物馆藏

泥质，褐色，质地坚硬，印面4字完整，背部有检痕。

“东平陵丞”封泥

Clay Impression of Seal

汉（前202—220）
济南市博物馆藏

“东平陵丞”封泥

Clay Impression of Seal

汉（前202—220）
济南市博物馆藏

“齐铁官丞”封泥

Clay Impression of Seal

西汉（前202—8）
济南市博物馆藏

印文为篆书“齐铁官丞”4字。铁官，秦代始置。据史料记载，汉代在产铁、产铜多的郡国设有铁官、铜官等官职。

半两钱范

Molds for Casting Banliang Bronze Coins

汉（前 202—220）
东平陵城遗址出土
济南市博物馆藏
长 27.5 厘米，宽 11 厘米

陶扑满

Taopuman (Pottery utensils used in ancient China to store bronze coins, similar to modern piggy banks)

汉（前 202—220）
东平陵城遗址出土
章丘区博物馆藏
通高 11.8 厘米，底径 7.5 厘米，腹径 13.5 厘米

扑满，即储钱的一种盛具，顶部开一狭口，可将零散铜钱投入其中。陶扑满只有入口，没有出口。钱装满后，则将其敲碎取之，“满则扑之”，故名“扑满”。

冶铁工场

东平陵为汉代铁官所在地，故城发现西汉时期大面积冶铁工场，显示出东平陵城作为汉代北方工业重镇的地位。西汉中期的双圈结构熔铁炉为首次发现，在冶金史上具有重要价值。遗址内发现的大量铁器，既有一般的铸铁和可锻铸铁，也有强韧性铸铁、碳钢和高碳工具钢，表明两千年前东平陵城的冶铁制造业已十分成熟和发达。《后汉书·郡国志》在描述济南国时称“东平陵，有铁”，济南国所辖的“历城，有铁”。铁，成为济南国的标志。

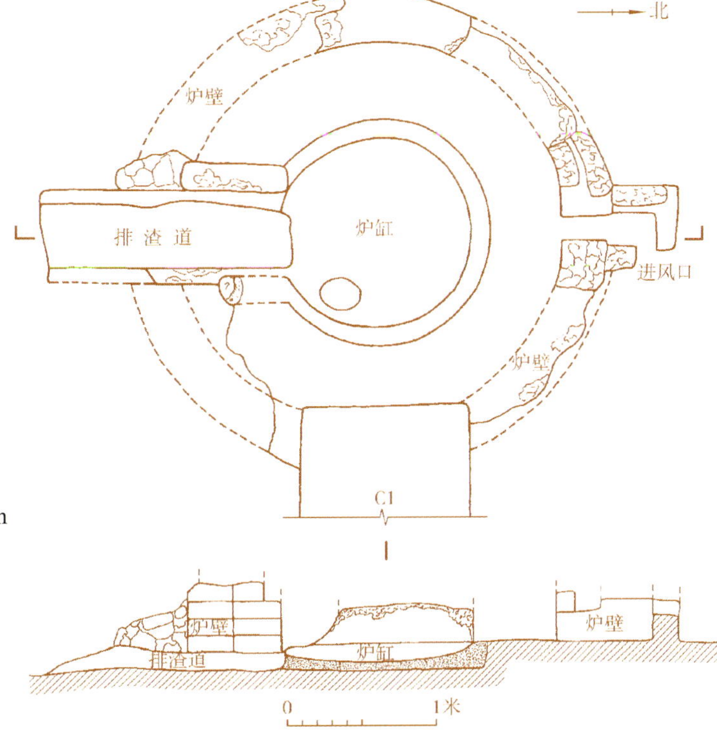

冶铁炉 L1 平、剖面图
Cross-section Diagram in Plan of Iron-smelting Furnace L1

Iron-smelting Workshop

Dongpingling was the place of the iron officials of the Han Dynasty. The discovery of large-scale iron-smelting workshops during the Western Han Dynasty in the ancient city shows the status of Dongpingling as a major industrial town in the northern region of the Han Dynasty. The double-ring structure melting furnace in the mid Western Han Dynasty was discovered for the first time and has important value in the history of metallurgy. The large number of iron artifacts discovered in the site include both general cast iron and malleable cast iron, as well as strong and tough cast iron, carbon steel and high carbon tool steel, indicating that the iron-smelting and manufacturing industry in Dongpingling City was very mature and developed two thousand years ago. When describing the State of Jinan, *the Book of the Later Han: Treatise on Prefectures and States* says, “There are irons in Dongpingling. There are irons in Licheng, a place under the jurisdiction of the State of Jinan.” Iron had become the symbol of the State of Jinan.

五齿铁钩

Five-clawed Iron Hook

汉（前 202—220）
东平陵城遗址出土
章丘区博物馆藏
残宽 12.1 厘米，高 1.2 厘米，
銎径长 3.8 厘米，宽 2.8 厘米

铁凿头

Iron Chisel Head

汉（前 202—220）
东平陵城遗址出土
章丘区博物馆藏
总长 18.0 厘米，刃宽 3.0 厘米

凿身呈“T”形，凿上端有一圆形凸起，凿下端呈长方形。整体器物锈蚀，用于生活生产。

铁剑

Iron Sword

汉（前 202—220）
东平陵城遗址出土
章丘区博物馆藏
长 121 厘米

兵器，剑身细长，前端尖锐，后有柄，两边锋刃。

圆孔铁锄

Round-hole Iron Hoe

汉（前 202—220）
东平陵城遗址出土
章丘区博物馆藏
残长 14.0 厘米，圆孔径 2.0 厘米

近椭圆片状，中间有孔。刃部稍残。

铁齿轮
Iron Gear

汉（前 202—220）
东平陵城遗址出土
章丘区博物馆藏
高 1.8 厘米，直径 5.8 厘米

汉代就有计算里程的记程鼓、测方向的指南车、记时器漏壶、运载工具木牛流马等机械，推断这些仪器均采用齿轮。这件铁齿轮的出土，对于研究我国古代机械工程的发展提供了重要的实物资料。

铁铧（huá）帽
Iron Ploughshare Cap

汉（前 202—220）
东平陵城遗址出土
章丘区博物馆藏
高 19.0 厘米，宽 29.5 厘米

农具。呈圆弧状三角形，用于耕田翻土作业。

济南国——危山汉墓兵马俑坑

汉文帝十六年（前 164）封刘辟光为济南王，以东平陵为济南国之都。位于章丘区圣井街道的危山汉墓，极有可能是西汉济南王刘辟光之墓。危山汉墓兵马俑坑是山东地区发现的第一座保存完好的兵马俑陪葬坑，是当时发现的我国继秦始皇兵马俑、陕西咸阳杨家洼兵马俑之后的第三大兵马俑坑。危山汉墓兵马俑坑南北长 9.7 米、宽 1.9 米、残深 0.8—0.9 米，出土陶俑 172 个、陶马 55 匹、陶车 4 辆、盾牌 60 余面、构件 150 余个，反映了汉代显贵出行的兵、车、马队列形式，塑造了骑兵、步兵和鼓乐队伍的形象。危山汉墓的考古发现入选 2003 年度全国十大考古新发现。

The State of Jinan: Terracotta Warriors Pit of the Han Tomb in Weishan

In the 16th year of Emperor Wen of the Han Dynasty (164 BC), Liu Biguang was enfeoffed as King of Jinan and Dongpingling was made the capital of Jinan. The Han Tomb in Weishan located in Shengjing Street, Zhangqiu District is most likely to be the Tomb of Liu Biguang, King of the State of Jinan during the Western Han Dynasty. The Terra Cotta Warriors Pit in the Han Tomb in Weishan is the first well preserved funerary pit of the Terra Cotta Warriors discovered in Shandong, and the third largest Terra Cotta Warriors Pit at that time in China after Qin Shihuang's Terra Cotta Warriors and Yangjiawa Terra Cotta Warriors in Xianyang, Shanxi. The Terra Cotta Warriors Pit in the Han Tomb in Weishan is 9.7m in length from north to south, 1.9m in width, and 0.8-0.9m in depth. 172 pottery figurines, 55 pottery horses, 4 pottery chariots, more than 60 shields and more than 150 components have been unearthed, reflecting the formation of soldiers, chariots and horses for the travel of dignitaries in the Han Dynasty, and shaping the image of cavalry, infantry, and drum band. The archaeological discoveries of the Han Tomb in Weishan had been selected as one of the top ten new archaeological discoveries in China in 2003.

危山汉墓兵马俑坑
Terracotta Warriors Pit of the Han Tomb in Weishan

危山汉墓兵马俑 1 号坑前排骑俑
Terracotta Warriors on Horseback in Front Row of Pit No.1 from Han Tombs in Weishan

西汉济南国王刘辟光墓

西汉景帝三年（前 154），济南王刘辟光因参与发动七国之乱，兵败自杀，济南废国，复改为郡。《章丘县志》载，平陵王墓在危山之巅。当地民间传说，汉景帝命人将刘辟光葬在危山。考古勘探显示，危山汉墓整体布局并不完整，兵败自杀的刘辟光之陵墓建设没有完工当在情理之中。从墓葬时代和规模看，危山汉墓基本可定为西汉济南国王刘辟光之墓。

Tomb of Liu Biguang, King of the State of Jinan in the Western Han Dynasty

In the third year of Emperor Jing of the Western Han Dynasty (154 BC), King Liu Biguang of the State of Jinan committed suicide after participating in the rebellion of the Seven States. Jinan was abolished and changed to a prefecture. As recorded in *Annals of Zhangqiu County*, the tomb of King of Pingling is on the top of Weishan. According to local folklore, Emperor Jing of Han ordered Liu Biguang to be buried on Weishan. Archaeological exploration has shown that the overall layout of the Han tomb in Weishan is incomplete; and it is reasonable that Liu Biguang, who committed suicide after being defeated, did not complete the construction of his tomb. Judging from burial period and scale, the Han tomb in Weishan can be basically determined as the tomb of Liu Biguang, the king of the State of Jinan during the Western Han Dynasty.

陶持盾俑

Shield-holding Pottery Figurine

西汉（前 202—8）
危山汉墓兵马俑坑出土
章丘区博物馆藏

泥质灰陶，手制而成，仅双手为模制后安装。站立状，头戴章甫冠，身着过膝长袍，通体施一层白衣，白衣表层施红色彩绘。俑的前部有盾牌，盾牌为多边形，断面呈三角形，后部有把手，绘粉红彩。

彩绘陶击鼓俑

Painted Pottery Figurines Beating the Drum

西汉（前 202—8）
危山汉墓兵马俑坑出土
章丘区博物馆藏
俑长 22.8 厘米，宽 13 厘米，高 41 厘米
鼓长 23 厘米，宽 22.3 厘米，高 45.5 厘米

在汉代，建鼓舞是盛行于宫廷和民间的一种舞蹈。该击鼓俑出土于危山汉墓一号坑，表现出击鼓状态，右手后摆，左手摆至胸前，造型生动，神态逼真，动感十足。建鼓有一方形的底座，上有一覆碗形座，楹柱贯穿鼓身，陶鼓四周各安置一个小鼓。古代作战通常进攻时击鼓、撤退时鸣金，击鼓起到鼓舞士气的重要作用。

陶骑马俑

Pottery Riding Figurines

西汉（前 202—8）
危山汉墓兵马俑坑出土
章丘区博物馆藏
马长 64.8 厘米，宽 18 厘米，高 58 厘米
俑长 23 厘米，宽 9.6 厘米，高 42.8 厘米

骑马俑，危山一号俑坑出土，泥质灰陶，手制而成，形体为蹲坐骑马状，头戴幅巾，面部凸鼓，通体施一层白衣，白衣表层施红色彩绘。

彩绘陶马

Painted Pottery Horse

西汉（前 202—8）
危山汉墓兵马俑坑出土
章丘区博物馆藏
长 62 厘米，宽 16 厘米，高 60 厘米

泥质灰陶，马身为手塑，马尾及耳均为模制后安装而成。马的形体呈站立状，昂首张口。通体施一层白衣，白衣表层马背上施红色菱形彩绘。

济北国——双乳山西汉济北王陵

济北国是楚汉及汉时的一个诸侯国，大致位于山东西北部。位于长清区归德街道的双乳山汉墓，应为西汉济北王刘宽之墓。该墓依山为陵，向下凿岩成穴，平面呈“甲”字形，总长85米，总深22米，总面积约1447平方米，总凿石量约8800立方米。如此巨大的工程量和规模，在已发掘的汉代诸侯王陵中居首位。葬具为三棺二椁，即五重棺椁制。棺均为漆棺，内红外黑，棺表漆饰云纹。关于汉代济北国的文献记载很少，双乳山汉墓的发现对于研究汉代济北国的历史及其地理位置均有重要意义。该项发掘入选1996年度全国十大考古新发现。

The State of Jibei: Mausoleum of the King of Jibei of the Western Han Dynasty in the Shuangru Mountain

The State of Jibei was a vassal state during the Chu-Han period and the Han Dynasty, roughly located in the northwest of Shandong Province. The Han Tomb in Shuangru Mountain located in Guide Street, Changqing District should be the tomb of Liu Kuan, the King of Jibei in the Western Han Dynasty. The tomb is based on the mountain as a mausoleum, and the rock is chiseled down into a cave. The tomb is in the shape of Chinese character “ 甲 ”, with a total length of 85m and a total depth of 22m. The total area is approximately 1,447m^2, and the total amount of stone chiseled is about 8,800m^3. Such a huge amount and scale of works rank first among those of the tombs of the feudal princes of the Han Dynasty that have been excavated. The burial assemblage comprises three inner coffins and two outer coffins, constituting a five-tiered coffin-chamber system. The coffins are all lacquered coffins, the inner surfaces are black, and the coffin surfaces are painted with cloud patterns. There are very few literature records about the State of Jibei in the Han Dynasty. The discovery of the Han Tomb in Shuangru Mountain is of great significance for studying the history and geographical location of the State of Jibei in the Han Dynasty. The excavation was selected as one of the top ten new archaeological discoveries in the country in 1996.

*全国重点文物保护单位
National Key Cultural Relics Protection Unit

双乳山汉墓俯视图
Top View of the Han Tomb in Shuangru Mountain

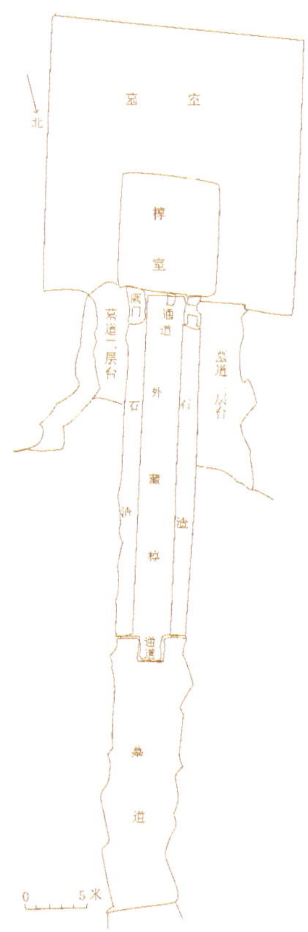

双乳山一号汉墓平面图
the Floor Plan of the Han Tomb No.1 in Shuangru Mountain

棺椁复原示意图
Schematic Diagram of the Restoration of Coffins

双乳山一号汉墓发掘现场
Excavation Site of Han Tomb No.1 in Shuangru Mountain

随葬器物

双乳山汉墓随葬铜器、玉器、车马器具等2000余件，主要出土于椁室和外藏椁中，其中有铜鼎9件，形制相近，大小依次递减，应为列鼎。内棺内墓主周围出土文物百余件，其中玉器50余件，计有玉覆面1套17件、玉枕1套9件、玉璧5件、九窍塞9件、猪形手握2件等。在死者的头骨下面，平铺着一层金饼，总共有20枚，大者直径6.5厘米左右，每枚重220克左右。金饼上多刻有文字，多数为“王”字，少数为“齐”或“齐王”。

Funerary Artifacts

There are more than 2,000 pieces of bronze, jade, carriage models or symbolic carriage and horse accessories buried in the Han Tomb of Shuangru Mountain, mainly unearthed in the inner coffins and the outer coffins. Among them, there are 9 bronze tripods, which are similar in shape and structure and decreasing in size. They should be Lie Ding (vessel displaying a variety of delicacies). More than 100 cultural relics were unearthed around the tomb owner inside the inner coffins. Specifically, there are more than 50 jade artifacts, including a set of Yu Fumian (jade mask for the deceased) with 17 components, a set of jade pillow consisting of 9 interlocking pieces, 5 Yu Bi (ritual discs), 9 body orifice plugs and 2 pig-shaped hand grips. Under the skull of the deceased, there is a layer of gold cake lying flat, with a total of 20 pieces, the largest of which has a diameter of approximately 6.5m and weighs approximately 220g each. The golden cakes are generally engraved with characters, most of which have inscriptions of “King”, and a few have the inscriptions of “King” or “King of the State of Qi”.

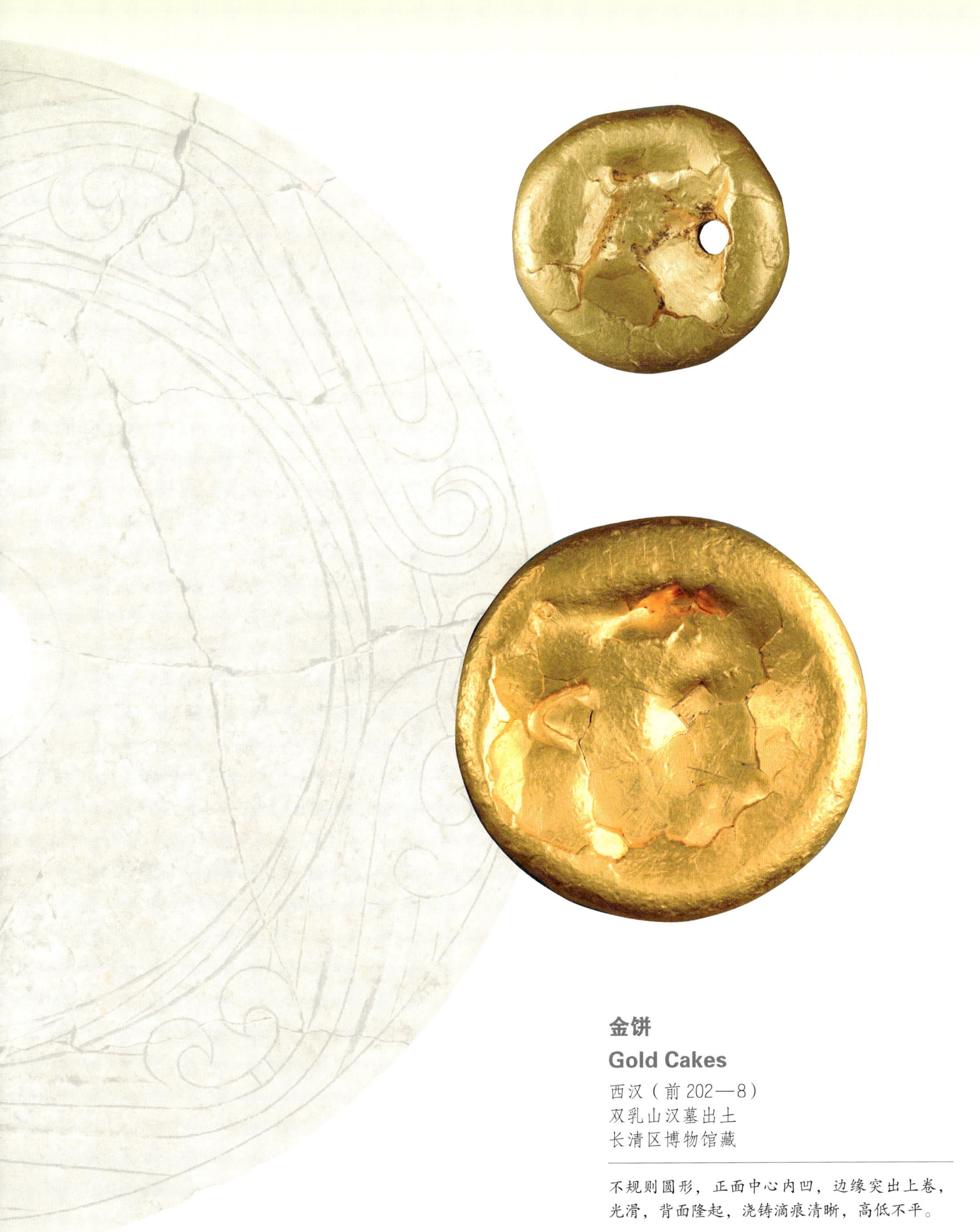

金饼

Gold Cakes

西汉（前 202—8）
双乳山汉墓出土
长清区博物馆藏

不规则圆形，正面中心内凹，边缘突出上卷，光滑，背面隆起，浇铸滴痕清晰，高低不平。

玉覆面

Yu Fumian (Jade Death-mask)

西汉（前 202—8）
双乳山汉墓出土
长清区博物馆藏
长 20 厘米，宽 23 厘米，厚 2.9 厘米

由额、颐、腮、颊、颌、颏、耳等 17 块玉片和鼻罩组合而成，形状为脸形，非常形象，左右对称，上下协调，部位恰当，浑然一体。眼睛、嘴巴由相对玉片对应磨出，并非独片相罩；各玉片内侧下棱和鼻罩边缘处斜穿细微孔，孔孔对应，相互缀连，连线似为丝缕之类；鼻罩采取透雕工艺，用一块玉石雕镂而成，形如半锥状，内琢空，鼻梁直挺，两翼微鼓，边缘外折，整体丰满盈溢。鼻罩两翼透雕云雷纹，下端透雕两个三角形鼻孔，余部外表线雕云雷纹，造型新颖，雕琢精细。除鼻罩外，其余玉片均为素面。

玉枕

Jade Pillow

西汉（前 202—8）
双乳山汉墓出土
长清区博物馆藏
长 42.1 厘米，宽 7.5 厘米，高 10.4 厘米

由玉片、玉板、玉虎头和竹板组合而成，玉件共计 14 件。枕的中心部分分为上、中、下三层，分别为枕面、枕蕊和枕底。枕的上侧两端各立一虎头，其下各接一玉板形足。多数玉件上都有纹饰，或透雕或浮雕或线雕而成，图案有云纹、几何纹、动物纹等，其中以玉虎头雕琢最为精致，虎头作竖鼻、睁目、斜胡须状。

玉耳塞
Jade Ear Plugs

玉鼻塞
Jade Nasal Plug

玉口唅
Jade Artifacts to be Held in the Mouth

玉阴罩
Jade Shadow Mask

玉肛塞
Jade Anal Plug

左玉手握
Left-Hand Jade Hand-Grip Artifact

右玉手握
Right-Hand Jade Hand-Grip Artifact

九窍塞

Nine Body Orifice Plugs

西汉（前 202—8）
双乳山汉墓出上
长清区博物馆藏

出土的九窍塞由玉口唅、玉耳塞、玉鼻塞、肛塞、玉阴罩和玉手握组成。其中，玉口唅做成蝉形放置于墓主人的口中，乃取“蝉蜕复生”之意。

随葬车马

墓内共发现马车 5 辆，其中真车 3 辆、小车 2 辆，同时发现大量鎏金车马明器。3 辆大车的形制、结构各不相同。1 号车为双辕车，驾 1 匹马，车长 3.9 米，配有伞盖。车马器中除车轴上的铁器，其余青铜器均镶嵌有错金银花纹。2 号车为单辕车，驾驷马，长约 4 米，宽 2.62 米，是 3 辆车中最为高贵的一辆。其车身装饰豪华，与《后汉书·舆服志》所载“王车”特征一致。3 号车也是独辕车，驾二马，车身长 4.1 米，车身饰有彩绘纹样，配有华盖，车马器也多为鎏金。

Funeral Artifacts of Chariots and Horses

The tomb complex yielded five carriages, comprising three functional full-scale carriages and two ceremonial miniature models, accompanied by substantial quantities of gilded carriage models or symbolic carriage and horse accessories. The shape and structure of the three large carriages are different. Carriage 1 features a double-shafted configuration, drawn by a single horse, with a total length of 3.9m and a structural canopy. Except for the ironware on the axles of the chariots and horses, all other bronze artifacts are inlaid with staggered gold and silver patterns. Carriage 2 features a single-shafted configuration, drawn by a 4 horses, with a total length of 2.62m. It is the most noble of the three carriages. The chariot exhibits lavish decorative schemes that demonstrate exact correspondence with the specifications of “Carriage of King” documented in the Treatise on Fashions and Costumes in the Book of the Later Han, Chapter 29. Carriage 3 features a single-shafted configuration, drawn by 2 horses, with a total length of 4.1m. The carriage’s body is decorated with painted patterns, equipped with a canopy, and the carriage models or symbolic carriage and horse accessories were mostly made of gilt bronze.

双辕车车舆马具名称说明图
Labeled Diagram of Harness Components for the Compartment of an Ancient Chinese Double-Shaft Horse Carriage

鎏金青铜盖弓帽

Gilt-Bronze Chariot Canopy Finials (Gai Gong Mao)

西汉（前 202—8）
双乳山汉墓出土
长清区博物馆藏

盖弓帽是汉代马车上用于固定和装饰车盖边缘的金属配件，主要起到固定车盖织物、防止滑脱的作用。

错金银青铜环

Bronze Ring with Gold and Silver Inlay

西汉（前 202—8）
双乳山汉墓出土
长清区博物馆藏

环的一面饰错金银图案，飞禽走兽被刻划得惟妙惟肖，线条虽细如发丝，但显得坚韧有力。

错金银青铜车軎（wèi）

Bronze Shaft Head Piece for Chariot with Gold and Silver Inlay

西汉（前 202—8）
双乳山汉墓出土
长清区博物馆藏

马车固定车轴两端的青铜构件，用于防止车轮脱落。

错金银青铜轴饰

Bronze Shaft Decoration with Gold and Silver Inlay

西汉（前 202—8）
双乳山汉墓出土
长清区博物馆藏

马车车轴上的装饰部件，左右各一，起着稳定伏兔（减震装置）、保护车毂（gǔ）、避免泥土沉积等作用。

济北王刘宽

西汉天汉四年（前 97），济北王刘胡去世，其子刘宽继承王位。后元二年（前 87），刘宽因被皇帝问罪“以刃自颈死”，济北国除。双乳山汉墓墓主颈下玉枕外侧放置两件有意破碎的玉剑璏（wèi），虽不能贸然断定是与墓主自刎有关，但这种现象极为罕见，应有其特殊寓意。

Liu Kuan, King of Jibei

In the fourth year of the Tianhan reign of the Western Han Dynasty (97 BC), King Liu Hu of Jibei passed away, and his son Liu Kuan inherited the kingship. In the second year of the Later Yuan of the Western Han Dynasty (87 BC), Liu Kuan accused by the Emperor,then he suicided upon hearing the news. Later, the State of Jibei was also abolished. Adjacent to the exterior of the jade headrest beneath the tomb occupant's neck at Han Tomb at the Shuangru Mountain, archaeologists discovered two intentionally fragmented jade sword slide fittings. Although the direct correlation with self-inflicted injury cannot be definitively established, this exceptionally rare phenomenon likely embodies specific ritual symbolism.

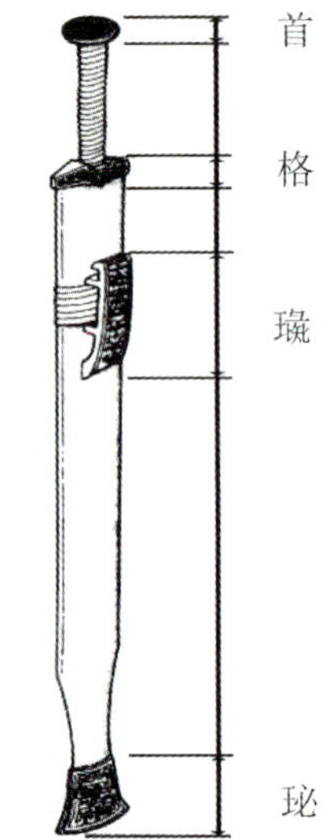

玉剑饰
Decorations of Jade Sword

玉剑璏

Jade Sword Slide Fittings

西汉（前 202—8）
双乳山汉墓出土
长清区博物馆藏
长 7.9 厘米，宽 1.8 厘米，厚 0.8 厘米

此西汉玉剑璏共 3 件。其中，1 件附在剑鞘上，2 件放置在墓主颈下玉枕外侧。浸为鸡骨白色，间有青绿色，两端下卷，背面有一长方形带穿，表面饰谷纹。

夫人私府——腊山汉墓

腊山汉墓位于济南市市中区腊山产业科技园内腊山东面的山坡上，它北距经十西路约 3000 米，东距二环西路约 2900 米。2001 年 7 月—9 月发掘，共出土陶器、铜器、铁器、水晶、玛瑙等遗物 70 余件，其中以鼎、罐、壶为主要组合的陶器占多数。

整个墓葬平面为折尺形，由墓道和墓室组成。墓室形制较特殊，大致分为前庭、前室、后室三部分。前庭南部与墓道相连，前庭东部与前室相连，前室东部与后室相连，三部分在宽度的变化上相差不大。根据墓葬形制及出土遗物，再结合出土的封泥和印章，可知此墓为西汉早期墓葬，墓主人应是一位列侯夫人。

Madam's Private Residence: The Han Tomb at Lashan

The Lashan Han Tomb is located on the eastern slope of the Lashan Industrial Science and Technology Park in Shizhong District, Jinan City. It is approximately 3,000m north of Jingshi West Road and approximately 2,900m east of Second Ring West Road. Excavated from July to September 2001, more than 70 relics mainly including pottery, bronze, iron, crystal and agate were unearthed, among which pottery composed of tripods, jars and pots accounted for the majority.

The plane of the entire tomb is in the shape of a folding ruler, consisting of a tomb passage and a tomb chamber. The tomb chamber has a unique shape and structure and is roughly divided into three parts, including the vestibule, the antechamber, and posterior chamber. The southern part of the vestibule is connected to the tomb passage; the eastern part of the vestibule is connected to the anterior chamber; and the eastern part of the anterior chamber is connected to the posterior chamber. There is not much difference in width between the three parts. Based on the shape of the tomb and the unearthed relics, combined with the excavated clay impression of seals and the seals, it can be inferred that this tomb is from the early Western Han Dynasty, and the owner of the tomb should be the wife of a Marquis.

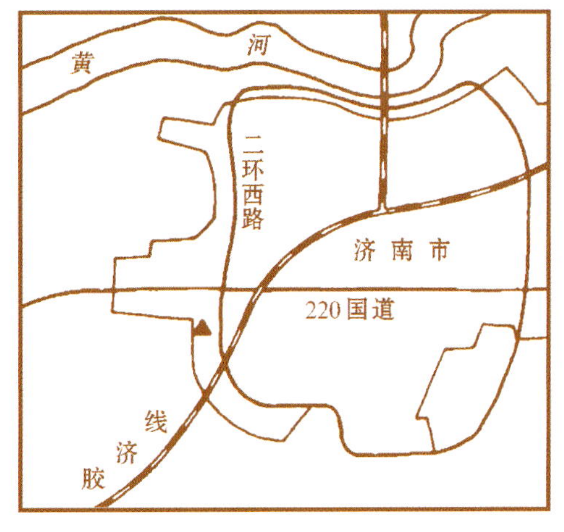

腊山汉墓位置示意图
Location map of the Han Tombs in Lashan

腊山汉墓全景
Panoramic View of Han Tombs in Lashan

七鼎之礼

腊山汉墓出土陶鼎7件，大小基本相同。带弧形盖，子母口。深腹，圜底，耳上端向外倾斜。器身腹部、器盖各彩绘一周三角纹。根据《公羊传·桓公二年》何休注曰："天子九鼎、诸侯七、大夫五、元士三也。"但是从春秋战国开始，用鼎制度有些混乱，出现了僭越现象。不但列国诸侯，甚至贵族也使用九鼎。如春秋晚期寿县蔡侯墓、战国曾侯乙墓、平山中山王墓都使用九鼎。西汉早期的墓葬仍保存着战国礼器制度的遗风，因此，此时的九鼎墓是皇帝或王室之墓，七鼎应是上大夫（卿）之墓。此墓正是袭用战国时期的诸侯上卿之礼。

The Tribute of Seven Tripods

Seven pottery tripods were unearthed from the Han Tomb in Lashan, basically the same size. These ritual tripods feature domed lids with tongue-and-groove joints. These ritual tripods exhibit a deep-bodied form with integrated hemispherical base casting. The handles demonstrate a characteristic outward inclination at the upper ends. The globular body's mid-section (belly) and domed lid each exhibit an encircling band of polychrome triangular patterns. *The Gongyang Commentary on the Spring and Autumn Annals - In the Second Year of Duke Huan*, as annotated by He Xiu of the Han Dynasty, it was prescribed hierarchical ritual vessel use system: "i.e., 9 tripods for the emperors, 7 tripods for the feudal, 5 tripods for senior officials, and 3 tripods for the junior officials." However, since the Spring and Autumn Period and the Warring States Period, the ritual system of using tripods had been somewhat chaotic, and there had been a phenomenon of usurpation. Not only the feudal lords of various states, but even the nobles used the nine tripods. For example, the tomb of Caihou in Shou County in the late Spring and Autumn Period, the tomb of Zeng Houyi in the Warring States Period, and the tomb of King Zhongshan in Pingshan all used nine tripods. The tombs of the early Western Han Dynasty still preserve the legacy of the Warring States ritual system, therefore, the Nine Tripod Tombs at that time were the tombs of emperors or royal families, and the Seven Tripod Tombs should be the tombs of high-ranking officials. This tomb is a tribute to the feudal lords and officials with high rank of the Warring States period.

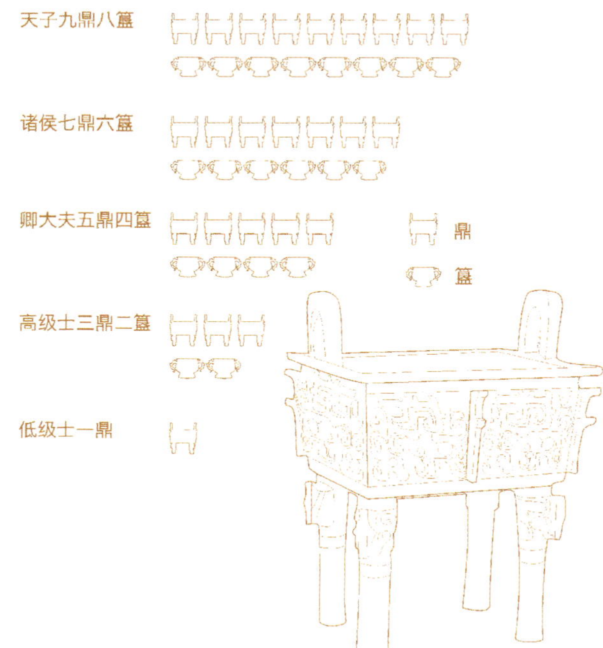

列鼎制度示意图
Schematic Diagram of the System for Lie Ding (Vessel Displaying a Variety of Delicacies)

“妾嫕”玉髓印章
Chalcedony Seal

西汉（前 202—8）
腊山汉墓出土
济南市考古研究院藏
高 1.4 厘米，边长 1.9 厘米

印章为玉髓质，印面呈方形，覆斗形纽，纽正中穿孔，篆书阴文“妾嫕”。

“傅嫕”水晶印章
Crystal Seal

西汉（前 202—8）
腊山汉墓出土
济南市考古研究院藏
高 1.7 厘米，边长 2.4 厘米

印章为水晶质，印面呈方形，覆斗形纽，纽正中穿孔，鸟虫书体阴文“傅嫕”。

龙形玉饰
Dragon-shaped Jade Ornaments

西汉（前 202—8）
腊山汉墓出土
济南市考古研究院藏
长 4.4 厘米，宽 0.8 厘米，高 0.3 厘米

龙形，片状，龙首较短小，吻前伸，略上翻，尾部残缺，阴线雕刻谷纹。

“夫人私府”封泥
Clay Impression of Seal

西汉（前 202—8）
腊山汉墓出土
济南市考古研究院藏
长 2.6 厘米，宽 2.6 厘米，厚 0.7 厘米

方形，上有“夫人私府”篆书印文。

东汉嗣侯——大觉寺村汉墓

大觉寺村汉墓出土了1000余块玉衣片，穿孔中残留有铜丝，推测墓主享受铜缕玉衣的葬制，身份应是东汉晚期的嗣侯（继承侯爵封号者）。该墓位于长清区归德街道大觉寺村附近，处于东汉济北国范围内。地表残存有高约1米的封土，墓室为大型砖石混筑结构，南北总长47米，东西最宽14米，由墓道、前室、东西耳室、中室和后室组成，墓中出土陶器、铜器、金银器等110余件。

Sihou (Heir to the Title of Marquis) of the Eastern Han Dynasty: Han Tomb in Dajue Temple Village

More than 1,000 jade garment pieces were unearthed from the Han tomb in Dajue Temple Village, with bronze wires remaining in the holes. It is speculated that the tomb owner enjoyed the burial system of copper thread jade clothing, and his identity should be that of an heir to the title of marquis in the late Eastern Han Dynasty. The tomb is located near Dajue Temple Village, Guide Street, Changqing District, within the scope of the State of Jibei in the Eastern Han Dynasty. There is still a mound of earth on the surface, which is about 1m in height. The tomb chamber is a large-scale structure made of bricks and stones. The total length from north to south is 47m, and the widest part from east to west is 14m. It is composed of a tomb passage, a front chamber, east and west side chambers, a middle chamber, and a back chamber. More than 110 artifacts, such as pottery, bronze wares, and gold and silver wares, were unearthed from the tomb.

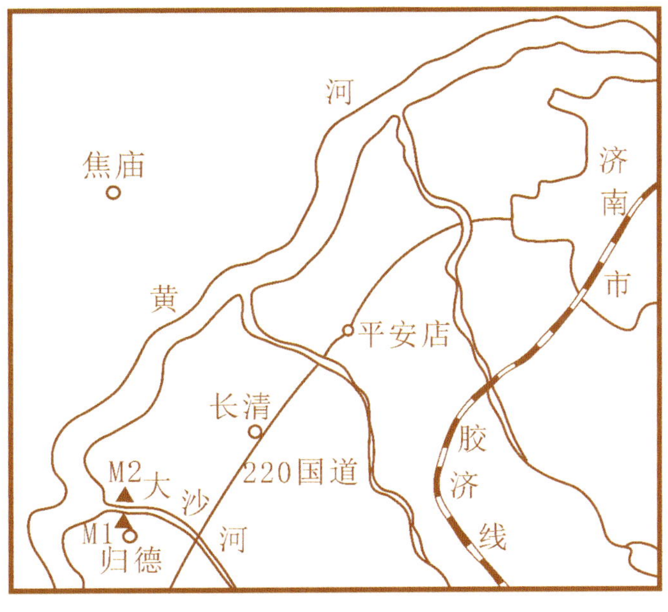

大觉寺汉墓位置示意图
Location Map of Han Tomb in Dajue Temple

大觉寺汉墓发掘现场
Excavation Site of Han Tomb in Dajue Temple

石兽形砚台

Stone beast-shaped Inkstone

东汉（25—220）
大觉寺汉墓 M2 出土
长清区博物馆藏
长 26 厘米，宽 9.2 厘米，高 25.2 厘米

该砚台分上下两合，整体呈猛兽载人形象。兽口大张，口内残有红色彩绘，颌下有须，全身覆鳞，左右各两翅膀，刻划出羽毛。脊上左右各两人背对而坐。右侧还有一兽，全身覆鳞。

陶楼

Pottery Miniature Building Discovered among Han Burial

东汉（25—220）
大觉寺汉墓 M2 出土
长清区博物馆藏
长 48 厘米，宽 24.5 厘米，高 32 厘米

该陶楼分上下两层。上层四面坡顶，檐下有4根立柱，前后各开一小窗。下层四面窗，正脊上左右各有一吻。底座除正面一侧，其余都有栏杆。

三层绿釉红陶楼

Three-storey Green Glazed Red Pottery Building

东汉（25—220）
大觉寺汉墓 M1 出土
长清区博物馆藏
长 36 厘米，宽 20 厘米，总高 72 厘米，底宽 15 厘米

共分3层。最上层为庑（wǔ）殿顶，正脊两端有角状吻，楼体正面有斜方格漏窗，两侧有挑出的斗拱支撑屋檐。中间一层与上层结构相同。最下面一层正面中间为两扇门。

高足鼓腹绿釉盘口红陶壶

Red Pottery Pot with High-footed Bulging Belly and Green Glazed Rim

东汉（25—220）
大觉寺汉墓 M1 出土
长清区博物馆藏
左：高 52 厘米，口径 19 厘米，底径 21 厘米
右：高 50 厘米，口径 20.5 厘米，腹径 26 厘米，底径 22 厘米

盘口，方唇，折沿，颈细长，鼓腹，圜底，大喇叭形足。

济水之南

第二章 济水之南

济南古城的发展演变

济南古城，其范围大致相当于明清时期的济南府城。千百年来，济南依泉而生、因泉而兴。考古证实，早在距今4000多年前的龙山文化时期，济南古城区域内就有人类活动，此后聚落和城市发展从未间断。“先有平陵城，后有济南府”的说法，是对济南区域中心变迁的概括。西晋末年，济南郡治从东平陵迁至历城。自此，现在的济南古城一带成为历代郡国、州府的行政中心。宋代以后，济南逐渐成为山东省的政治、经济和文化中心，至明代正式成为山东首府，并逐渐形成较为稳定的城市布局。

Chapter II. South of the Ji River: The Development and Evolution of the Ancient City of Jinan

The ancient city of Jinan roughly covers an area corresponding to the Jinan Prefectural City during the Ming and Qing dynasties. For thousands of years, Jinan has been born and prospered by springs. Archaeological evidence has shown that there were human activities in the area of the ancient city of Jinan as early as the Longshan Culture period more than 4,000 years ago. Since then, the settlement and urban development have never been interrupted. The saying that “Pingling City came first, followed by Jinan Prefecture” is a summary of the transformations in the regional center of Jinan. At the end of the Western Jin Dynasty, the ancient city area of present-day Jinan had historically served as the administrative cores for successive commanderies, principalities, prefectures, and provincial administrations of the past dynasties. After the Song Dynasty, Jinan gradually became the political, economic and cultural center of Shandong Province, and officially became the provincial capital of Shandong in the Ming Dynasty, forming a relatively stable urban layout.

泺畔古邑

济南古城地区自龙山文化时期便有人类活动的踪迹。早在商代，趵突泉附近已有先民生活。东周时期，齐国在济南设邑。此后，济南从泺水之滨的小邑逐步发展为郡国、州府的治所。汉代，济南历城是全国 49 处铁官所在地之一。据文献记载，西晋时期，济南郡治西迁，在古历城以东另建新城，形成东、西二城的格局；至唐代，筑城活动使东、西二城融为一体，城市规模再次扩大，经济、文化更趋繁荣。

泉水聚落（史前—汉代）

魏晋以前，历城城址位于今济南古城西部，夹处泺水和历水之间的“历内（汭）”一带。考古工作者在济南古城的中西部发现了史前和商代的文化遗存。春秋时期，齐国在此设历下邑，将其作为齐国西陲的军事重镇，由此工商业也得到一定的发展。汉代，历城的手工业发达，中央政府在此设立铁官。

Luo-Riverside Ancient Settlement

There had been traces of human activities in the ancient city area of Jinan since the Longshan Culture period. As early as the Shang Dynasty, there were ancients living near Baotu Spring. During the Eastern Zhou Dynasty, the State of Qi established a regional seat in Jinan. Since then, Jinan gradually developed from a small settlement along the ancient Luo River to administrative cores of commanderies, principalities, prefectures, and provincial administrations. During the Han Dynasty, Licheng in Jinan was one of the 49 places where iron officials were located in China. According to literature records, during the Western Jin Dynasty, the county seat of Jinan was relocated to the west and a new city was built to the east of ancient Licheng, forming a pattern of two cities in the east and west. By the Tang Dynasty, the fortification activities had integrated the east and west cities of Jinan, expanding the city’s size and promoting economic and cultural prosperity.

Spring-nourished Settlements (From Prehistoric Times to the Han Dynasty)

Before the Wei and Jin dynasties, the site of Licheng was located in the western part of the ancient city of Jinan, sandwiched between the Luoshui River and the Lishui river in the area of “Li Nei (the confluence of rivers)”. Archaeologists have discovered cultural relics from prehistoric and the Shang Dynasy in the central western part of the ancient city of Jinan. During the Spring and Autumn Period, the state of Qi established fortified settlement of Lixia (in present-day Jinan) as a strategic military garrison on its western frontier, while commercial and industrial activities flourished moderately. In the Han Dynasty, the handicraft industry in Licheng was developed, and the central government established iron officials here.

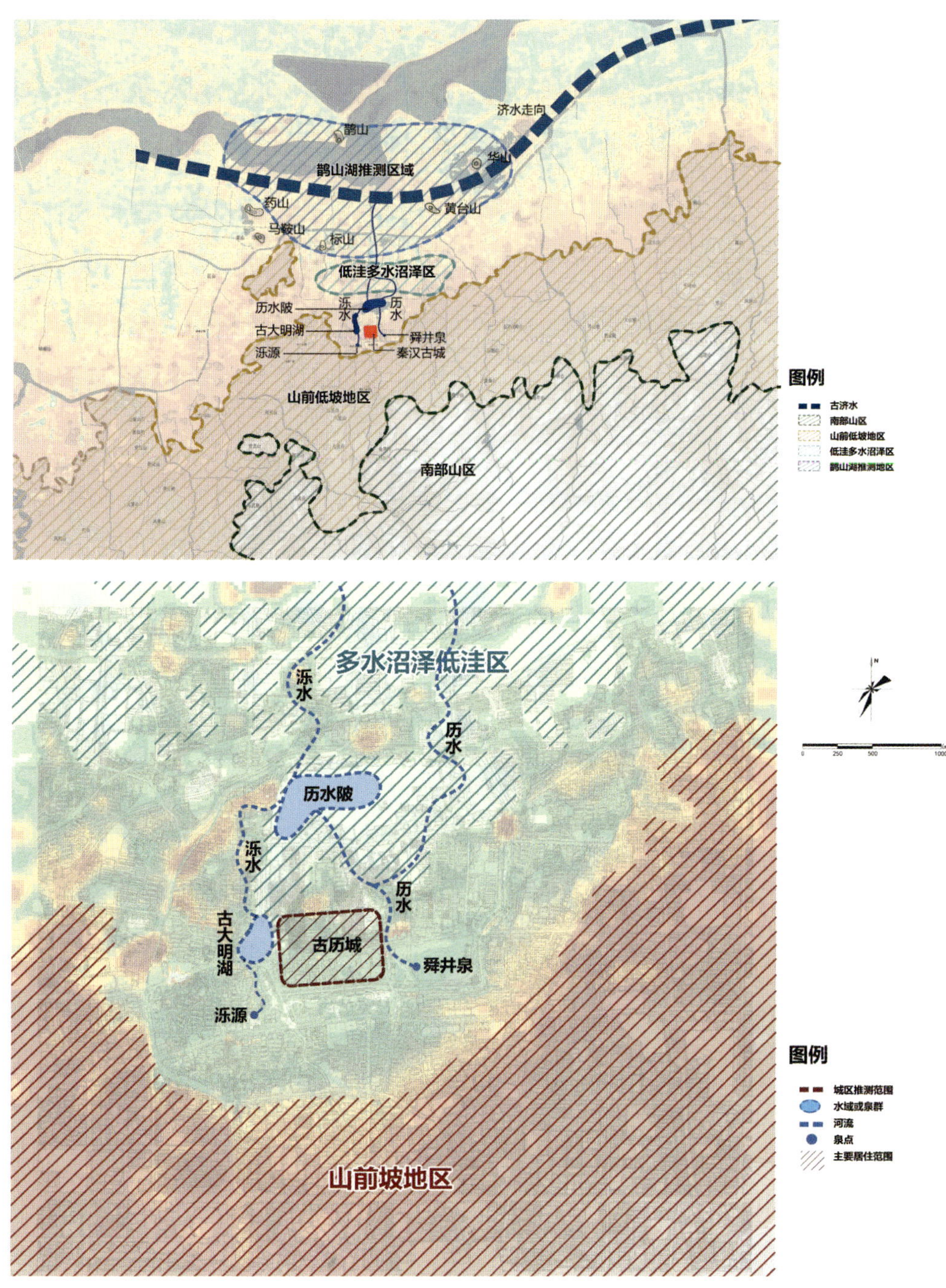

魏晋之前的城市和水环境
The urban and water environment before the Wei and Jin Dynasties

城西的河流名叫泺水，发源于趵突泉，向北注入济水；城东的河流名叫历水，发源于舜井泉群，北流转西汇入泺水。

古城 4000 年——天地坛街遗址

天地坛街遗址是济南第一次在古城区域范围内发现的龙山文化遗存，这一发现把古城区内有人类活动的历史向前推进至距今 4000 多年前。该遗址位于古城的中南部，考古发掘揭露出一个龙山文化时期的灰坑及 21 个战国至汉代的灰坑，出土大量生活用器及板瓦、筒瓦等建筑构件，显示出济南古城地区人类生活的延续性。

The 4000-Year-Old Ancient City: Tianditan Street Historic Site

Historic Site of Tianditan Street is the first discovery of Longshan Culture relics within the ancient city of Jinan, which advances the history of human activities in the ancient city area to more than 4000 years ago. The historic site is located in the central southern part of the ancient city. Archaeological excavations have discovered an ash pit from the Longshan Culture period and 21 ash pits from the Warring States to the Han Dynasty. A large number of household utensils, as well as building components such as plate tiles and semicircle-shaped tiles, have been unearthed, demonstrating the continuity of human life in the ancient city area of Jinan.

天地坛街遗址发掘全景
Excavation Panorama of Historic Site of Tianditan Street

红陶鬲（lì）足

Red Pottery Cooking Tripod with Hollow Legs (Red Pottery Li Pot)

龙山文化时期（距今约 4500—4000）
天地坛街遗址出土
济南市考古研究院藏
长 8.8 厘米，最大直径 4.5 厘米

近圆锥形，足尖圆滑。

天地坛街名称的由来

明朝时期，天地坛街位于德王府的正南方，街的左右分别兴建了山川坛和社稷坛，用以祭祀天地，天地坛街因此而得名。清代，山川坛和社稷坛被废除，随后在其遗址上兴建了许多民居和铺房，但街名保留了下来。

Origin of the Name of Tianditan Street

During the Ming Dynasty, Tianditan Street was located directly south of Prince De's Mansion. On both sides of the street, Shanchuan Altar and Sheji Altar were built to offer sacrifices to heaven and earth, hence Tianditan Street got its name. In the Qing Dynasty, the Shanchuan Altar and the Sheji Altar were abolished, and subsequently many residential and commercial buildings were built on their historic sites, but the name of this street was preserved.

商族居邑——旧军门巷遗址

早在3000年前，古城西南部的旧军门巷一带已有先民生活。旧军门巷遗址发现了商代中期的5件陶鬲足和2件陶甗（yǎn）足，它们在陶质、形制上与古城周边的历城大辛庄遗址出土的陶鬲、陶甗颇为接近。

Shang Clan Settlement Site: Historic Site of Former Junmen Lane

As early as 3,000 years ago, the area around Former Junmen Lane in the southwest of the ancient city had already been inhabited by the ancients. Five red pottery cooking tripods and two pottery cooking utensils with legs from the mid Shang Dynasty were discovered at Former Junmen Lane. They are quite similar in pottery quality and shape to the pottery cooking tripods (Pottery Li Pot) and pottery cooking utensils with legs (pottery Yan) unearthed at Daxinzhuang in Licheng around the ancient city.

旧军门巷遗址发掘全景
Panoramic View of the Excavation of Historic Site of Former Junmen Lane

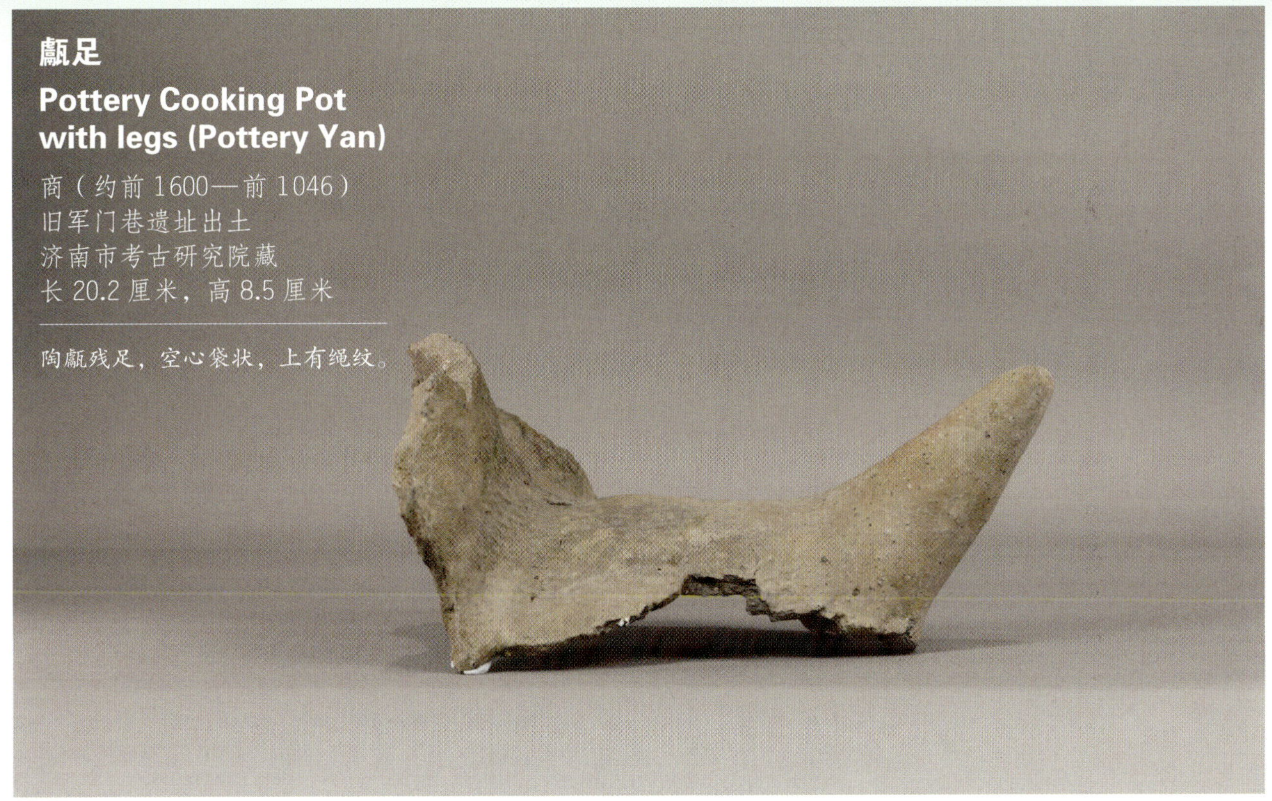

甗足
Pottery Cooking Pot with legs (Pottery Yan)
商（约前 1600—前 1046）
旧军门巷遗址出土
济南市考古研究院藏
长 20.2 厘米，高 8.5 厘米

陶甗残足，空心袋状，上有绳纹。

旧军门巷名称的由来

旧军门巷的名称与明代设立的山东巡抚衙门有关。成化元年（1465），此街西侧建巡抚衙门。随着巡抚职责的增加，其衙门逐渐改称督抚军门，此街因此得名军门巷，清末改称旧军门巷。

Origin of the Name of Former Junmen Lane

The name of Former Junmen Lane is related to the Imperial Commissioner's Yamen in Shandong established in the Ming Dynasty. In the first year of Chenghua (1465), the Imperial Commissioner's Yamen was built on the west side of this street. With the increase of the duties of the imperial commissioner, his Yamen gradually changed its name to Junmen of the imperial commissioner, hence the street was named Junmen Alley. In the late Qing Dynasty, it was renamed Former Junmen Lane.

千年货贝——中银大厦战国墓

早期商业以物易物，海贝是最早的货币。位于历下区泺源大街东段的中银大厦战国墓一次出土骨贝币600余枚，是战国时期济南经济和商贸发展的见证。墓中出土的贝币，多数为单孔，也有两孔贝。贝上有孔，主要是便于将它们组合成串。贝币质地坚固，不易损坏，携带方便，既可以分离为一个个单体用于小额支付，又可以组合成串用于购买大宗商品，成为早期最为流通的货币之一。

Thousands-year Bone Coins Imitating the Shape of Shells: The Warring States Tomb at the Bank of China Tower

Early commerce involved bartering, with seashells being the earliest form of currency. More than 600 bone coins imitating the shape of shells were unearthed from the Warring States Tomb of the Bank of China Tower located in the eastern section of Luoyuan Avenue, Lixia District, which is a witness to the economic and commercial development of Jinan during the Warring States period. The shell coins unearthed from the tomb are mostly single-hole shells, but there are also two-hole shells. There are holes on the shells, mainly to facilitate their combination into strings. Shell coins have sturdy textures, are not easily damaged, and are easy to carry. They can be separated into individual units for small payments, or combined into strings for purchasing bulk commodities, which became one of the most circulating currencies in the early days.

骨贝币

Bone Coins Imitating the Shape of Shells

战国（前476—前221）
中银大厦战国墓出土
济南市考古研究院藏

兽骨雕刻，仿货贝，形状扁平，正面中间贯穿凹槽，两侧有细密刻划，背面有双孔。

齐陶量

Pottery Measuring Instrument of the Qi

战国（前 476—前 221）
天桥区委大院防空干道工程出土
济南市博物馆藏
口径 21.2 厘米，底径 20 厘米，腹围 71.5 厘米

夹砂灰陶，圆筒形，敛口，深腹，平底。腹下部戳印一字，应释为“市”或“粜（tiào）”。该器为战国晚期齐国陶量，经测容积为 4220 毫升。其制作规整，又戳印铭文，甚少见。

铜提梁壶

Bronze High-handled Pot

战国（前 476—前 221）
中银大厦战国墓出土
济南市考古研究院藏
高 49.5 厘米，口径 11.3 厘米，底径 11.2 厘米

弧形盖顶，侈口，束颈，鼓腹，寰底，圈足。腹部和盖部附加环形铁环，以提梁相连。

燕明弧背刀币

Arc-backed Ming Knife Coin of Yan (Warring States period bronze currency with curved blade profile)

战国（前 476—前 221）
济南柴油机厂出土
济南市博物馆藏
通长 14 厘米，通宽 2.1 厘米，通高 0.2 厘米

凹刃，背微弧，刀首略宽，呈斜坡状，多数有外廓，身背边廓与柄廓相连，但明显高于柄廓。窄柄，刀柄正，背面各有纵纹二道，少数刀柄背面二纵纹伸入刀身一小部分。刀环一般作圆形或椭圆形。制作粗糙，含铅量较高。

齐法化刀币

Fa Hua Knife Coin of Qi (Three-character Knife Coin cast by the Qi State during the Warring States period)

战国（前 476—前 221）
济南市博物馆藏
通长 18.5 厘米，通宽 2.7 厘米，通高 0.4 厘米

刀身刻“齐法化”三字。“齐”即齐国，“法化”意为官方认证的标准货币。齐法化刀币是齐国货币制度成熟的标志，表明齐国在一定程度上实现了境内货币的统一。

历城铁官——按察司街、运署街汉代冶铁遗址

汉代在全国设 49 处铁官，其中济南郡在东平陵和历城设有铁官。古城东部按察司街、运署街一带发现的大规模汉代冶铁遗址，可能就是历城铁官下辖的铁工场址。遗址出土了一批冶铁用的陶范、炼炉碎片、铁矿石、铁渣、铁器等与冶铁活动有关的遗物。此外，遗址内发现了多处汉代水井，表明充沛的泉水资源影响着冶铁工场的分布。

Licheng Iron Official: Iron: Smelting Historic Site at Anchasi Street and Yunshu Street of the Han Dynasty

During the Han Dynasty, there were 49 iron officials throughout the country, among which Jinan Prefecture established iron officials in Dongpingling and Licheng. The large-scale iron smelting historic sites of the Han Dynasty discovered at Anchasi Street and Yunshu Street in the eastern part of the ancient city may be the site of the iron-melting workshop under the jurisdiction of the Licheng Iron Officials. A number of pottery molds, furnace fragments, iron ore, iron slag, iron implements and other relics related to iron-smelting activities were unearthed at the site. A number of water wells of the Han Dynasty have been discovered in the site, indicating that abundant spring water resources affect the distribution of iron-smelting workshops.

按察司街、运署街名称的由来

按察司街，得名于明清时期设在此街的提刑按察使司衙门。“按察司”是元、明、清时期主管一省司法事务的官署。

运署街，因清代都转盐运使司署设立在此而得名。“都转盐运使司”是元、明、清时期掌管食盐产销的机构。

Origin of the Names of Anchasi Street and Yunshu Street

The name of Anchasi Street comes from the Surveillance Commission Office, which was established on this street during the Ming and Qing dynasties. The Surveillance Commission Office was the local authority in charge of judicial affairs in a province during the Yuan, Ming and Qing dynasties.

Yunshu Street was named after the establishment of Duzhuan Salt Transport Commission Office during the Qing Dynasty. Duzhuan Salt Transport Commission Office was the agency in charge of salt production and marketing in the Yuan, Ming and Qing dynasties.

运署街遗址考古现场
The Archaeological Site of the Historic Site of Yunshu Street

运署街遗址发现多眼古井
Multiple Ancient Wells Discovered in the Historic Site of Yunshu Street

汉代陶炼炉
Pottery-refined Furnace of the Han Dynasty

铁片

Iron Sheet

汉（前 202—220）
按察司街遗址出土
济南市考古研究院藏
长 13.2 厘米，宽 10 厘米

出土于汉代冶炼遗址，是研究当时济南地区手工业发展、农业器具与社会结构的重要实物资料。

六角锄陶范

Pottery Model of Hexagonal Hoe

汉（前 202—220）
运署街遗址出土
济南市考古研究院藏
长 25.8 厘米，宽 23.5 厘米，厚 6.5 厘米

灰陶质，为铸造铁质六角锄的模具。

墓葬艺术——无影山西汉墓地

无影山汉墓出土的陶俑表明，西汉早期的造型艺术已进入一个更高的阶段。其中，杂技俑是首次发现的西汉时代百戏的立体形象，生动反映了当时百戏艺术所达到的高度。无影山西汉墓地位于济南市区北部，共发现墓葬29座，清理14座，其中11号墓出土了一组乐舞、杂技、宴饮陶俑。陶俑共21个，固定在长67厘米、宽47.5厘米的陶盘上。表演者位于陶盘中心，其身后是乐队，两侧则是观众。

Tomb Art : The Western Han Tombs in Shadowless Mountain

The pottery figurines unearthed from the Han Tombs in Shadowless Mountain indicate that the plastic arts of early Western Han Dynasty entered a higher stage. Specifically, acrobatic figurines are the first three-dimensional images of the Hundred Plays of the Western Han Dynasty discovered, vividly reflecting the level of artistic achievement of the Hundred Plays at that time. The Western Han Tombs in Shadowless Mountain is located in the northern part of Jinan city, with a total of 29 tombs discovered and 14 cleared. Among them, a group of pottery figurines for music, dance, acrobatics and banquets were unearthed from Tomb 11. There are a total of 21 pottery figurines, fixed on pottery plates 67cm in length and 47.5cm in width. The performers is located in the center of the pottery dish, with the band behind them and the audiences on either side.

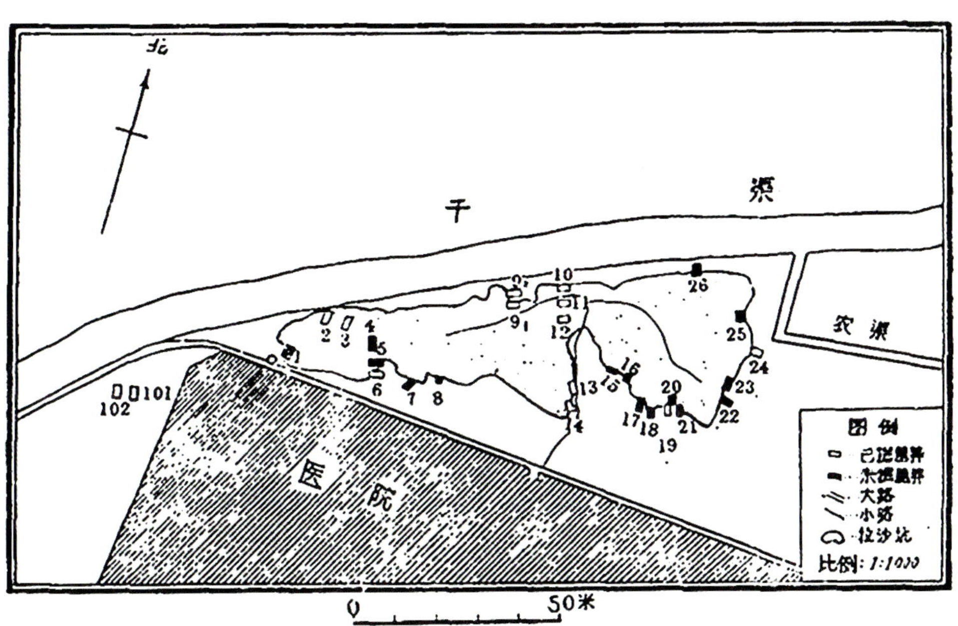

无影山西汉墓地分布示意图
Schematic Diagram of the Distribution of The Western Han Tombs in Shadowless Mountain

彩绘载人载鼎陶鸟

Color-Painted Human-and-Ding-Bearing Pottery Bird

西汉（前 202—8）
无影山西汉墓出土
济南市博物馆藏
长 45 厘米，宽 37.5 厘米，高 50.5 厘米，底座长 29 厘米，底座宽 18.5 厘米

泥质灰陶。鸟头小，胸凸，昂首短喙，颈与前胸绘鳞状羽纹，双翼左右平展，长尾微翘，两足踏长方形底座之上。鸟背塑有 3 人，后面一人着赭衣，双手撑圆盖伞，前面两人拱手相对，均着朱衣。鸟左右两翼各载一圆形盖鼎，鼎足为人形。陶鸟展翅欲飞，气势雄伟，似传说中的鸠鸟。鸠为吉祥之鸟，载美食满鼎，祝愿长者长生不老。整器造型新颖，构思巧妙，寓意深长。

彩绘负壶陶鸟

Color-Painted Pottery Bird with Pot

西汉（前 202—8）
无影山西汉墓出土
济南市博物馆藏
长 44 厘米，宽 38 厘米，高 52.5 厘米，底盘长 26.7 厘米，底盘宽 17.6 厘米

鸟形似鸠，双翼各载一壶，壶盖上绘朱色花纹，并饰 3 只朱色鸟头形扁纽。鸠鸟被喻为“不噎之鸟”，双壶中盛满稀粥，备老人饮用，可见汉代敬老爱老之风盛行。整器形态生动逼真，气势古拙凝重，从器型到寓意均具独到之处。

彩绘乐舞杂技陶俑

Painted Pottery Figurines of Music, Dance and Acrobatics

西汉（前 202—8）
无影山西汉墓出土
济南市博物馆藏

无影山汉墓出土的陶俑表明，西汉早期造型艺术进入一个更高的发展阶段。11 号墓出土的一组乐舞、杂技、宴饮陶俑塑造于长方形平座上，共计 21 个，表演者位于中心，后面是乐队，两侧是观众。其中，杂技俑是首次发现的西汉时代百戏的立体形象，生动反映了当时杂技艺术所达到的水平和高度。

汉代陶塑

汉代陶塑以其极高的艺术价值和重要的历史价值著称于世，是了解汉代社会风貌的重要材料之一。由于西汉中晚期后艺术形态逐渐向世俗化演变，以民间风貌为主要题材的陶塑艺术为其增添了浓厚的生活气息和乡土色彩。济南地区大量汉代陶塑的出土，为研究两汉时期济南人的生活提供了鲜活的材料。

Pottery Sculptures in the Han Dynasty

Han Dynasty pottery sculptures are renowned for their high artistic and historical value, and are one of the important materials for understanding the social landscape of the Han Dynasty. Due to the gradual evolution of artistic forms towards secularization after the mid to late Western Han Dynasty, pottery sculpture art with folk style as its main theme has added a strong atmosphere of life and local colors to it. The excavation of a large number of pottery sculptures of the Han Dynasty in Jinan provides vivid materials for studying the lives of Jinan people during the Eastern and Western Han dynasties.

绿釉陶井

东汉（25—220）
奥体中路汉墓出土
济南市考古研究院藏
宽 18.2 厘米，高 46.5 厘米

绿釉红陶。井身呈口小底大圆筒状，外塑缠绕枝花蔓，架顶塑一展翅飞鸟。

汉画像石

汉画像石是汉代厚葬之风盛行的社会背景下，出现的一种“大象其生以送其死”的丧葬艺术。其题材丰富，是政治制度、社会关系、生产能力、宗教信仰、道德观念及艺术水平等各方面内容的综合体现，具有很高的史料价值和审美价值。

孝堂山汉石祠位于长清区孝里铺村，是一座双开间单檐悬山顶房屋式建筑，也是我国现存的保存于地面最早的房屋式建筑。石祠内壁有精美的阴刻汉画像，内容包括车马出行、历史故事、神话传说、杂耍舞技、屠宰庖厨、汉胡交战、日月星辰、合围狩猎等题材，展现了汉代社会的风貌和人们的信仰追求，堪称汉代艺术的瑰宝。

Stone Relief of the Han Dynasty

Stone relief of the Han Dynasty was a funeral art form that was popular during the Han Dynasty, where the tradition of elaborate funeral was prevalent. The art of “The Grand Mimetic Representation of Life to Escort the Dead” was extremely popular during the Han Dynasty. Its rich themes were a comprehensive embodiment of various aspects such as political system, social relations, production capacity, religious beliefs, moral concepts, and artistic level, with high historical and aesthetic value.

The Han Stone Temple on Xiaotang Mountain is located in Xiaolipu Village, Changqing District. It is a double-room single-eave overhanging hilltop house-style building and also the earliest surviving house style building preserved on the ground in China. There are exquisite engraved Han portraits on the inner wall of the stone temple, covering topics such as travel with carriage and horse, historical stories, myths and legends, acrobatics and dance skills, slaughter and cooking, conflicts between the Han Dynasty and northern nomadic peoples, the sun, moon and stars, encirclement hunting, showing the style of Han society and people’s religious pursuits. It can be regarded as a treasure of Han art.

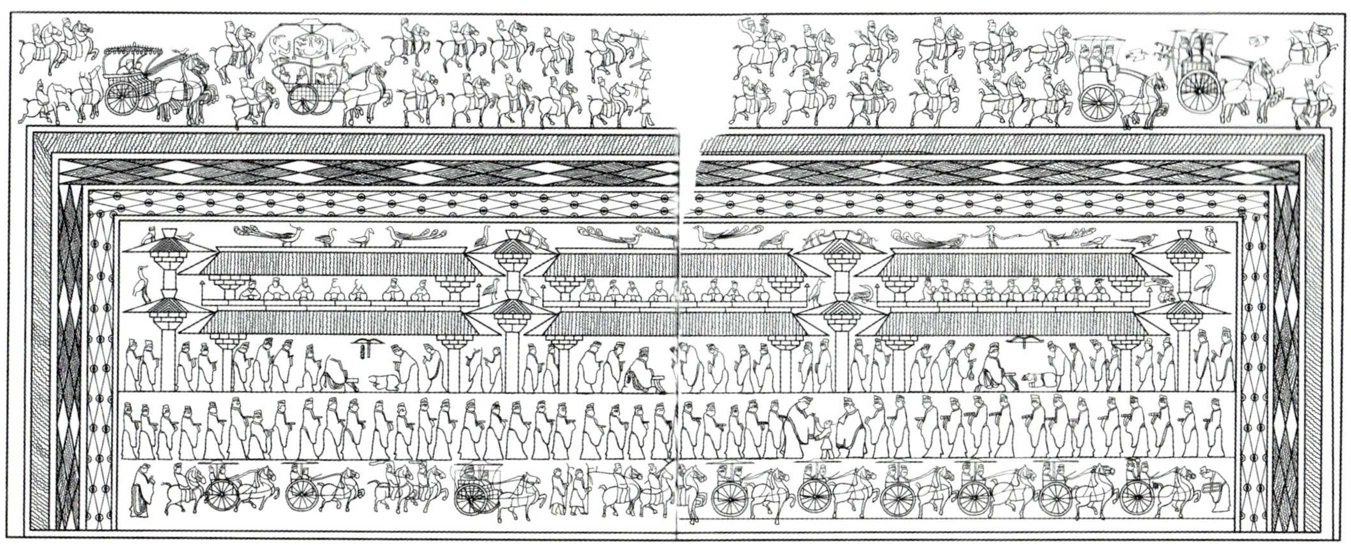

孝堂山汉石祠北壁画像摹本
Rubbing Copy of the Portraits on the North Wall of the Han Stone Temple on Xiaotang Mountain

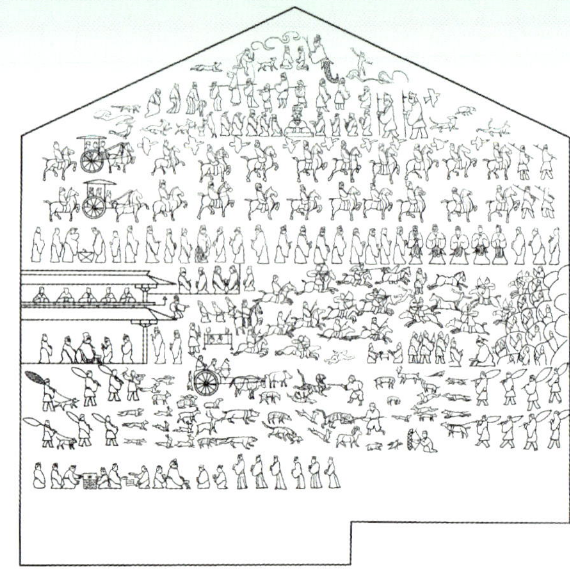

孝堂山汉石祠西壁画像摹本
Rubbing Copy of the portraits on the West Wall of the Han Stone Temple on Xiaotang Mountain

城郊墓地——魏家庄墓群

魏家庄墓群东距古城约 1 公里，应为城内居民的墓地，共发现战国至明清时期墓葬 168 座，其中汉墓 96 座。汉墓多为南北向，头向北，规模较小，一般长 2.5—3.8 米、宽 1.3—2 米，形制可分为土坑竖穴墓和土坑竖穴砖（石）椁墓，其中前者占半数以上。共出土陶、铁、铜、玉、石器等各类随葬品 630 余件（组），数量之多说明当时此地已是一重镇，特别是一次性出土汉代铁器 40 余件，乃全国罕见。

Suburban Cemetery: Weijiazhuang Tombs

Weijiazhuang Tombs were located approximately 1km east of the ancient city and are believed to be the cemetery of the residents of the city. A total of 168 tombs from the Warring States period to the Ming and Qing dynasties have been discovered, including 96 tombs of the Han Dynasty. Han tombs are mostly oriented in a north-south direction, with the head facing north, and are relatively small in scale. They are generally 2.5-3.8m in length and 1.3-2ms in width. The shapes and structures of tombs can be divided into vertical earth-pit tombs and vertical brick (stone) coffin tombs, of which the former accounts for more than half. More than 630 pieces (groups) of various burial objects such as earthenware, iron, bronze, jade and stone artifacts were unearthed, indicating that Jinan was a major city at that time. Especially, more than 40 iron artifacts of the Han Dynasty were unearthed at one time, which is rare in the country.

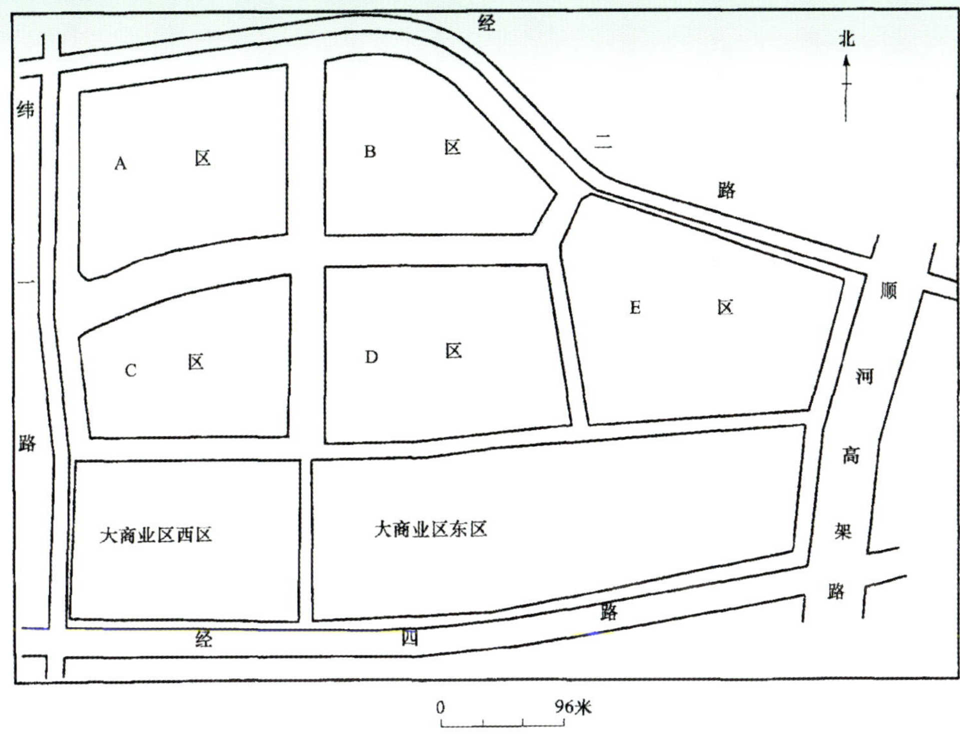

魏家庄遗址发掘区范围
Scope of Excavation Area of Historic Site of Weijiazhuang

家族墓地

魏家庄墓群内的墓葬分布具有一定的规律性，根据其分布区域和密集程度的不同，可将墓地划分为若干墓区。不同墓区的墓葬在形制、方向、随葬品组合、器物特征等方面多有相似或相同之处，说明不同墓区可能是分属不同家族的茔地。从出土印章分析，该地至少包含但氏和另一姓氏的家族墓地。

Family Cemetery

The distribution of the tombs in the Weijiazhuang Tombs have a certain regularity, and according to their distribution areas and density, the cemetery can be divided into several tomb areas. The tombs in different burial areas have similarities or commonalities in terms of shape, orientation, combination of burial objects and characteristics of artifacts, indicating that different burial areas may belong to different families. According to the analysis of the unearthed seals, the site contains at least the family cemeteries of Dan and another surname.

“但贤之印”铜印章

Bronze Seal of “Danxian”

西汉（前 202—8）
魏家庄墓地出土
济南市考古研究院藏
边长 1.25 厘米，高 1.13 厘米

印章为铜质，矮覆斗形印体，龟钮，阴文篆书“但贤之印”。

“但防之印”铜印章

Bronze Seal of “Danfang”

西汉（前 202—8）
魏家庄墓地出土
济南市考古研究院藏
边长 1.6 厘米，高 2.5 厘米

印章为铜质，方形印体，桥形钮，阴文篆书“但防之印”。

“但穫（huò）之印”铜印章

Bronze Seal of “Dan Huo”

西汉（前 202—8）
魏家庄墓地出土
济南市考古研究院藏
边长 1.72 厘米，高 1.71 厘米

印章为铜质，方形印体，桥形钮，阴文篆书“但穫之印”。

“但诵之印”铜印章

Bronze Seal of “Dansong”

西汉（前 202—8）
魏家庄墓地出土
济南市考古研究院藏
长 1.58 厘米，宽 1.55 厘米，高 1.67 厘米

印章为铜质，方形印体，龟形钮，阴文篆书“但诵之印”。

人形铜席镇

Human-shape Bronze Xizhen

西汉（前 202—8）
魏家庄墓地出土
济南市考古研究院藏
其一：高 9.4 厘米，长 8.3 厘米，宽 7.8 厘米
其二：高 9.5 厘米，长 9.3 厘米，宽 7.8 厘米
其三：高 8.1 厘米，长 9.4 厘米，宽 6.3 厘米
其四：高 7.6 厘米，长 7.5 厘米，宽 6.8 厘米

汉代人们通常席地而坐，压席子四角的镇成为日常的生活用具。席镇以鹿、虎、熊等动物造型居多，而人物造型少见。人形青铜镇表现的多为俳优艺人说唱的形象。俳优，是指以滑稽的动作、诙谐的言语娱乐他人的艺人。

铜鐎（jiāo）壶

Bronze Jiao Pot (Bronze Pot for Mixing and Warming Wine)

西汉（前 202—8）
魏家庄墓地出土
济南市考古研究院藏
长 26.8 厘米，宽 20.6 厘米，高 17 厘米

温酒器。带盖，矮直口，扁圆腹，三蹄形足，一侧鸟首形流，另一侧附一长直柄，柄断面为方形。

铜鉶（hóu）镂

Bronze Hou Lou (Bronze Pot with the combination of lid, pot body and handle)

西汉（前 202—8）
魏家庄墓地出土
济南市考古研究院藏
高 18.5 厘米，宽 17 厘米

食器，类似于现代的饭盒，用于盛放食物。

铜鋞（xíng）

Bronze Xing (Bronze Wine Pot)

西汉（前 202—8）
魏家庄墓地出土
济南市考古研究院藏
口径 12.5 厘米，高 20.3 厘米，最宽处 14.1 厘米

铜鋞是西汉时期常见的盛酒器或盛食器。其小口径、深筒结构利于保温，提梁便于携带。

立鹤踏龟铜博山炉

Bronze Boshan Incense Burner with Crane Standing on Turtle

西汉（前 202—8）
魏家庄墓地出土
济南市考古研究院藏
高 22 厘米，直径 23.6 厘米，底径 7.6 厘米

由盖、身、柄和座四部分组成。盖呈高耸山形，饰层峦叠嶂，中有诸多镂孔。柄为一昂首、展翅站立朱雀。座为一玄武，作龟昂首匍匐状，有蛇缠绕于龟身两侧。

堂狼洗

Tang Lang Xi (Bronze Washing Vessel)

东汉永元九年（97）
广智院街出土
济南市博物馆藏
通高 21.6 厘米，口径 44.6 厘米，腹围 125 厘米，底径 25 厘米

器内底有8字铭文“永元九年堂狼造作”。堂狼为汉代地名，位于西南地区犍为郡。

双子之城（西晋一隋唐）

据文献记载，西晋永嘉年间（307—313）因气候干旱导致河流断流，加之战乱破坏，济南郡治从东平陵西迁至河流、泉水丰沛的历城。从此，今济南市区成为历代郡国、州府的行政中心。刘宋元嘉九年（432）在济南郡侨置冀州，济南为州、郡两级治所。北魏改侨冀州为齐州。西晋至隋唐时期，文化多元，各民族融合发展。随着城市等级的提高和社会经济的发展，济南的城市规模有所扩大，形成东、西两城，并出现城郭。地处南北交通之地的济南，也在这一过程中迎来了新的繁荣景象。

Twin Cities (From Western Jin Dynasty to Sui and Tang Dynasties)

According to literature records, In the Yongjia period of the Western Jin Dynasty (307-313), rivers cutoff due to drought and chaos caused by wars, the county town of Jinan was relocated from Dongpingling to Licheng where there were had abundant rivers and springs. Since then, the urban area of present-day Jinan became the administrative cores for successive commanderies, principalities, prefectures, and provincial administrations throughout imperial Chinese history. In the ninth year of the Yuanjia reign of the Liu Song Dynasty (432), Ji Prefecture was established in Jinan commandery, with Jinan serving as the administrative center for both the levels of the prefectures and the provincial administration. During the Northern Wei Dynasty, Ji Prefecture was renamed Qi Prefecture. From the Western Jin Dynasty to the Sui and Tang dynasties, there was cultural diversity and the integration and development of various ethnic groups. With the improvement of urban level and the development of social economy, the urban scale of Jinan expanded, forming two cities in the east and west, and the city walls were built. Jinan, located in the north-south transportation hub, had also ushered in a new prosperous scene in this process.

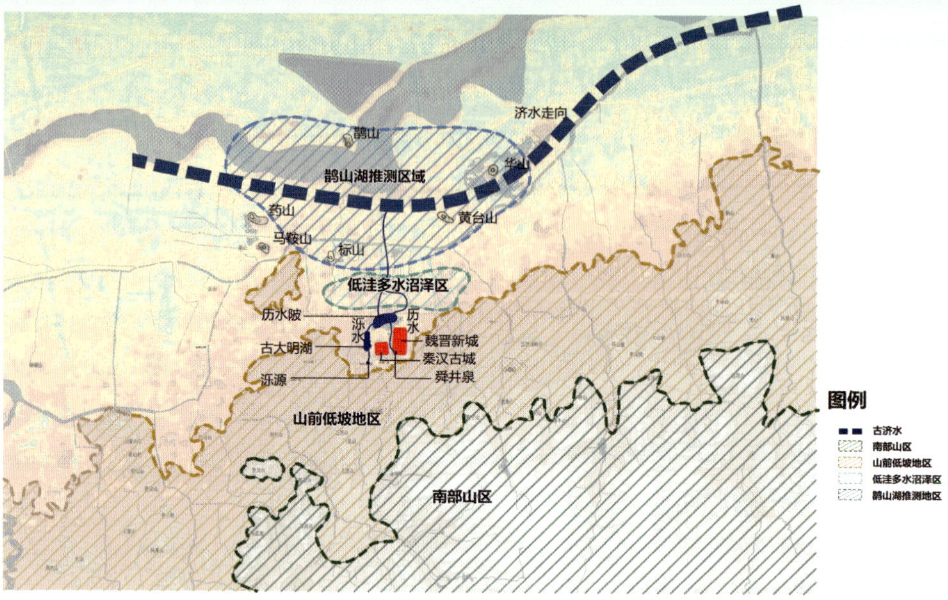

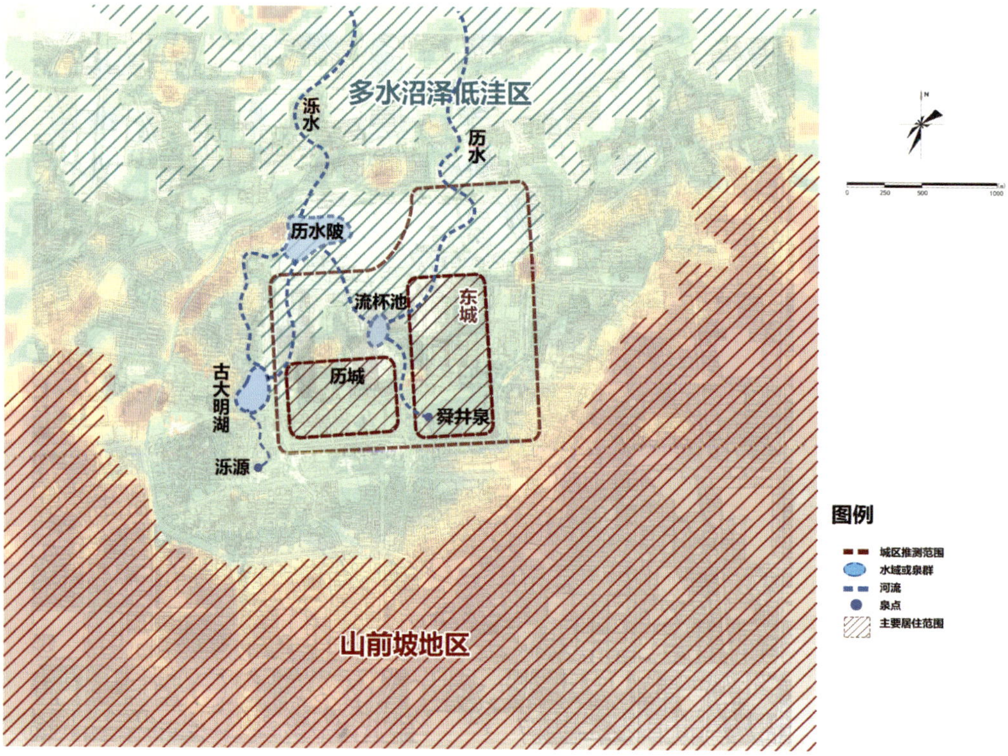

魏晋时期的城市和水环境

Urban and Water Environment during the Wei and Jin Dynasties

永嘉末年，济南郡治并没有迁入历城旧城，而是在古历城东侧扩大城垣，另建一座东城为郡治，济南城出现了东、西双子城的格局。《水经注》所勾勒出的北魏济南城的大致轮廓表明，东、西两城的范围已包括今济南老城厢的大部分区域。根据“筑城以卫君，造郭以守民”的原则，两座子城内应以官衙和驻军为主，而外郭则是工商业者等平民百姓的居住区。

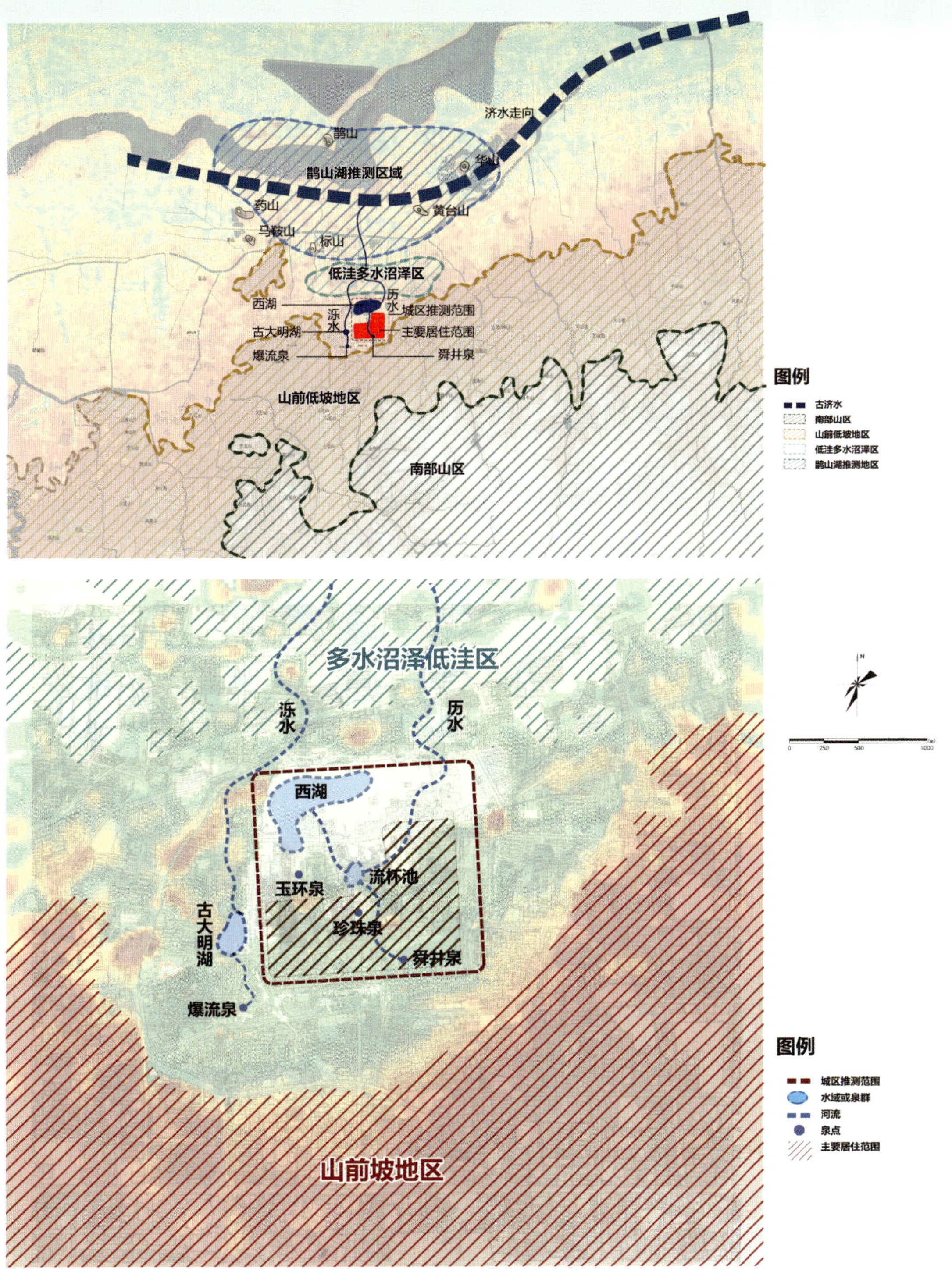

唐代的城市和水环境
Urban and Water Environment in the Tang Dynasty

唐代，筑城活动使东、西二城融为一体（一说为宋金元时期），城内诸泉汇入泺水的水道被城墙隔断，泉水泄流不及，遂于城西北地势低洼地带积水形成西湖（今大明湖），城内居住空间被挤占，洪涝时有发生。

晋代衣冠——历城区西晋砖室墓

历城区唐冶街道发现的一座西晋砖室墓，反映出当时中原文化因素对济南地区的影响。该墓平面近“甲”字形，由墓道、墓门、甬道、墓室构成。墓葬出土了一批具有典型时代特征的器物。其随葬器物组合与中原地区同期墓葬基本一致，而出土的圆形斜壁多子槅，又具有鲜明的地方特色。

Cenotaph in the Jin Dynasty: Brick Chamber Tomb of the Western Jin Dynasty in Licheng District

A brick-chambered tomb of Western Jin Dynasty was discovered in Tangye Street, Licheng District, reflects the influence of Central Plains cultural factors on the Jinan area at that time. The tomb has a nearly Chinese Character “甲”, consisting of the tomb passage, the tomb door, the corridor, and the tomb chamber. A number of artifacts with typical characteristics of the era were unearthed from the tomb. The burial assemblage shows essential congruence with contemporaneous tombs in the Central Plains region, while the excavated round multi-compartment box with sloping walls demonstrates distinct local characteristics.

历城区西晋砖室墓全景
Panoramic view of the Brick-chambered Tomb of the Western Jin Dynasty in Licheng District

陶多子槅（gé）

Pottery Multi Compartment

西晋（265—316）
历城唐冶永隆街西段西晋砖室墓出土
济南市考古研究院藏
高 6.8 厘米，口径 28.8 厘米，底径 19.3 厘米

圆形，敞口，平底，中央一圆格，内分 3 格，外圈分为 5 格。

北朝士族

东魏崔令姿墓

清河崔氏是魏晋至隋唐时期的著名大族。崔令姿墓发现于济南东郊圣佛寺村。据墓志铭记载，崔令姿是清河武城人，三国魏中尉崔琰（yǎn）的后人，东魏征北将军、右光禄大夫邓恭伯的夫人，其父为清河太守崔延伯。崔令姿去世于武泰元年（528），享年 29 岁。崔令姿墓志铭具有极高的艺术价值，展现了南北朝时期书法的风格特点。

Aristocrats of the Northern Dynasty

Tomb of Cui Lingzi in Eastern Wei Dynasty

The Cui family in Qinghe was a famous clan during the Wei, Jin, Sui, and Tang dynasties. The tomb of Cui Lingzi was discovered in Shengfosi Village in the eastern suburb of Jinan. According to the epitaph, Cui Lingzi was a native of Wucheng, Qinghe. She was a descendant of Cui Yan, a lieutenant of the Wei Dynasty during the Three Kingdoms period. She was the wife of Deng Gongbo, a general for the Northern Expedition of the Eastern Wei Dynasty and Right Imperial Minister of State. Her father was Cui Yanbo, the Taishou (prefectural-level governor) of Qinghe. Cui Lingzi passed away in the first year of Wutai (528) at the age of 29. The epitaph of Cui Lingzi has extremely high artistic value, showcasing the stylistic characteristics of calligraphy during the Northern and Southern Dynasties period.

崔令姿墓志铭

Epitaph Rubbings of Cui Lingzi's Tomb

东魏（534—550）
圣佛寺村崔令姿墓出土
济南市博物馆藏
碑长 47 厘米，碑宽 46.3 厘米，碑厚 7.8 厘米

志文为楷书，记载了墓主人的家世、生平以及后人对逝者的颂扬。志文为魏晋常见之方笔，结构趋于规整平实，是隶书向唐楷过渡的典范。

方髻男立像

Portrait of a Man in a Square Bun

东魏（534—550）
圣佛寺村崔令姿墓出土
济南市博物馆藏
长 8 厘米，宽 5.9 厘米，高 27.2 厘米

石像面露微笑，眉目舒展，着交领左衽（rèn）宽袖长袍。从发髻和装束上，可窥东魏时期社会世俗生活。

双髻女立像

Portrait of a Woman in Double Buns

东魏（534—550）
圣佛寺村崔令姿墓出土
济南市博物馆藏
长 6.7 厘米，宽 3.6 厘米，高 21 厘米

石像脸庞丰润，神情自如，着交领右衽宽袖长袍。

圆形石磨盘
Circular Stone Grinding Disc

东魏（534—550）
圣佛寺村崔令姿墓出土
济南市博物馆藏

磨盘分上下两部分，使用时下磨盘固定不动，旋转上磨盘以达到研磨效果。此磨盘为明器，不具实用性，但可反映出东魏时期磨盘的基本形制和农作物加工的情况。

石座狗
Stone-seat Dog

东魏（534—550）
圣佛寺村崔令姿墓出土
济南市博物馆藏

石狗呈蹲坐状，尖嘴，两耳向后贴于头上，下有一石座，二者系一块石头雕出。

北齐宜阳国太妃傅华墓

傅华乃北齐宰相赵彦深之母。赵彦深自幼丧父，由其母傅氏抚育长大，官至北齐宰相，封宜阳郡王。傅华以其子贵，封宜阳国太妃。傅华墓位于槐荫区段店后周王庄腊山河畔，出土 3 尊体量较大的石俑、10 余件陶俑，是山东地区出土北齐墓中形制最大、规格最高的墓葬。

Tomb of Fu Hua, Imperial Concubine of Yiyang State in Northern Qi Dynasty

Fu Hua was the mother of Zhao Yanshen, the prime minister of the Northern Qi Dynasty. Zhao Yanshen lost his father at a young age and was raised by his mother, Lady Fu. Zhao Yanshen rose to the position of prime minister of the Northern Qi Dynasty and was enfeoffed as the Prince of Yiyang. Fu Hua was honored by her son and was enfeoffed as the Empress Dowager of Yiyang. The tomb of Fu Hua is located on the banks of the Lashan River in Houzhouwangzhuang Village, Duandian Town, Huaiyin District. Three large stone figurines and over ten pottery figurines have been unearthed. It is the largest and highest-specification tomb of the Northern Qi Dynasty unearthed in Shandong Province.

傅华墓出土石俑
Stone Figurines Unearthed from Fu Hua's Tomb

傅华墓志志盖
Cover of the Epigraph of Fu Hua's Tomb

傅华墓志志石拓片
Epitaph Stone Rubbings of Fu Hua's Tomb

石俑

Stone Figurine

北齐（550—577）
腊山分洪河北齐宜阳国王太妃傅华墓出土
济南市考古研究院藏
肩宽 20 厘米，高 93.5 厘米；底座高 9 厘米

头左偏，戴帽盔，身穿开领左衽窄袖外衣，双手握于胸前。

石俑

Stone Figurine

北齐（550—577）
腊山分洪河北齐宜阳国王太妃傅华墓出土
济南市考古研究院藏
高 106 厘米，头长 20.7 厘米，宽 31.7 厘米；底座长 29 厘米，宽 21.6 厘米，高 10.7 厘米

身躯魁伟，披甲胄（zhòu），手持长兵（木制已腐朽），侧目相向。面目严肃，威风而立。

皇亲国戚——隋代吕道贵兄弟墓

吕道贵是隋文帝杨坚之舅。吕道贵墓是济南市发现的第一座隋代皇亲墓，对研究隋代历史、吕氏家族和隋代葬制具有重要价值。吕道贵兄弟墓位于市中区经八路纬四路。一号墓破坏严重，形制不明。二号墓为石筑双室墓，发现墓志 1 方，刻“仪同三司济南郡守吕道贵墓铭”，出土武士俑 2 件，石刻门楣 1 块。

Princes and Princesses of the Royal Family:Tomb of Lv Daogui and His Brother in the Sui Dynasty

Lv Daogui was the uncle of Yang Jian, Emperor Wen of Sui. The tomb of Lv Daogui is the first imperial tomb of the Sui Dynasty discovered in Jinan City, and has important value for studying the history of the Sui Dynasty, the Lv Family and the burial system of the Sui Dynasty. The Tomb of Lv Daogui and His Brother is located on Jingba Road and Weisi Road in Central District of Jinan. Tomb 1 was severely damaged and its shape and structure is unknown. Tomb 2 is a stone built double-chambered tomb. One side of the tomb epitaph was found with the inscription “Lv Daogui, Governor of Jinan Commandery with the Ceremonial Rank Equivalent to the Three Excellencies”. Two warrior figurines and one stone carved lintel were unearthed.

吕道贵墓志拓片

Rubbings of Lv Daogui's Epitaph

隋（581—618）
汇苑家园隋代济南郡守吕道贵墓出土
济南市考古研究院藏
宽 57 厘米，高 57 厘米，厚 15 厘米

青石质，志文为楷书，共 23 行，满行 23 字，首行书“仪同三司济南郡守吕道贵墓铭”。

武士镇墓石俑

Stone Figurines Guarding the Tomb

隋（581—618）
汇苑家园隋代济南郡守吕道贵墓出土
济南市考古研究院藏
高 79 厘米

青石制。面部丰满光洁，头戴高冠，身穿交领大袖长袍，双手拄宝剑于胸前，面目严肃，威风而立。

敕送舍利

舍利函是佛教安放高僧火化后尸骨和生前器物的容器。隋文帝曾赐30舍利于全国30州同建舍利宝塔，据《法苑珠林》记载，赐予齐州的舍利则“敕令法赞送舍利于泰山神通寺”。1973年初，四门塔修缮时，在塔心柱离地面高1.6米处发现一舍利石函，石函内放一铜函，铜函中藏有水晶珠4颗、黄琉璃珠7颗及绿琉璃珠9颗等。该舍利函可能就是当年由隋文帝敕送齐州供养的。

四门塔内顶部的三角石梁上，放有较大块的石板，在维修时发现其中一块石板反面的左下方有两行竖刻字迹，内容为“大业七年造”（正书，文字直径2—4厘米）。这一镌刻浅凹雕的发现明确了四门塔的建造年代，是为神通寺文物之重大发现之一。

Edict of sending the Sarira

A sarira reliquary is a Buddhist container used to enshrine the cremated remains and lifetime artifacts of venerated monks. Emperor Wen of the Sui Dynasty once granted 30 Sariras to build the Relics pagodas in 30 states of the country. According to the records in the Fayuan Zhulin, the Buddhist relics granted to Qi Prefecture was “a royal decree to send the sarira to Shentong Temple of the Mount Tai.” During the renovation of the Four-Gate Pagoda in early 1973, a stone sarira reliquary was discovered at a height of 1.6m above the ground on the central pillar of the pagoda. Inside the box was a bronze box containing 4 crystal beads, 7 yellow glass beads, and 9 green glass beads. The stone sarira reliquary may be the one that had been sent by Emperor Wen of Sui to Qi Prefecture for offerings.

On the triangular stone beam at the top of the Four-gate Pagoda, there are relatively large stone slabs. During maintenance, it was discovered that there are two vertical engraved lines on the lower left side of the reverse side of one of the stone slabs, which read “Made in the Seventh Year of Daye” (regular script , with a diameter of 2-4 cm). The discovery of this shallow concave carving clarifies the construction year of the Four-gate Pagoda and is one of the major discoveries of cultural relics in Shentong Temple.

铜函

Bronze Box

隋大业七年（611）
四门塔出土
历城区博物馆藏
通长 9 厘米，宽 9 厘米，通高 13 厘米，重 2 千克

铜函为盝顶铜函，边缘纹饰已模糊不清，函盖呈梯形，内空。

石函

Stone Box

隋（581—618）
长 31 厘米，宽 31 厘米，
高 30 厘米，重 52.23 千克

五铢钱

Wuzhu Coin

隋（581—618）
直径 2.2 厘米，厚 0.1 厘米

隋代五铢钱为方孔圆钱，“五”字交笔弯曲，“朱”字上笔方折。

半两钱

Banliang Coin

隋（581—618）
直径 2.7 厘米，厚 0.1 厘米

隋代半两钱为圆形方孔青铜钱，已锈成黑色。

骨指环

Bone Ring

隋（581—618）
直径 2.2 厘米，厚 0.34 厘米

骨指环为磨制指环，体形厚实而小，横截面呈半圆形。

银指环

Silver Ring

隋（581—618）
直径 2 厘米，厚 0.1 厘米

银指环为细银条盘旋而成，素面。

料珠

Glass Beads

隋（581—618）

料珠为圆鼓形，中心对钻一圆孔，其中绿色 9 个、黄色 7 个、白色 4 个。

名门望族——唐代朱满墓

乐陵朱氏是北朝时期中原地区一个很有名望的士族，他们本在乐陵，北朝后期一分支迁到青州。唐代朱满墓是济南首次发现的乐陵朱氏墓，位于历城区唐王街道的樊家遗址内，墓葬总体呈“刀把”型，通长约 10 米，由墓道、甬道和方形砖砌墓室组成。朱满墓随葬 50 余件唐三彩器，是山东地区出土唐三彩器数量最多、造型最全、制作最精美的一次，其中最重要的天王俑和镇墓兽均是盛唐时期比较典型的器物。同时，此墓还出土白瓷器 20 余件，墓志一合。据墓志记载，该墓埋葬时间为开元三年（715）。

Tomb of Zhu Man (A Notable and Great Clan) in the Tang Dynasty

The Zhu Clan in Laoling was a highly respected aristocratic family in the Central Plains during the Northern Dynasty. They originally lived in Laoling and later moved to Qing Prefecture as a branch in the later period of the Northern Dynasty. The tomb of Zhu Man in the Tang Dynasty is the first tomb of the Zhu clan of Leling discovered in Jinan. It is located in the Fanjia Site in Tangwang Street, Licheng District. The tomb is generally in the shape of a “knife”, with a total length of about 10m. It consists of a tomb passage, a corridor and a square brick-chambered tomb. More than 50 Tang tri-color glazed ceramics artifacts were buried in Zhu Man’s tomb. It is the discovery with the largest number, most complete shape and most exquisite production of Tang tri-color glazed ceramics artifacts unearthed in Shandong. Among them, the most important Heavenly kings figurines and tomb-guarding beasts are typical artifacts During the prosperous Tang Dynasty. At the same time, more than 20 pieces of white porcelain artifacts and epitaphs were unearthed from this tomb. According to the epitaph, the burial time of the tomb was in the third year of Kaiyuan (715).

朱满墓的墓室局部
Part of the Tomb of Zhu Man

三彩天王俑

Three-color Heavenly Kings Figurine

唐（618—907）
樊家遗址朱满墓出土
济南市考古研究院藏
高 101 厘米，底座长 21 厘米，底座宽 12.5 厘米，最宽 40 厘米

方脸，八字胡须，兜鍪（móu）上装饰展翅朱雀，肩覆披膊作龙首状。胸前左右各一护甲，左手握拳上举，右手叉于腰际，腰带之下左右各垂膝裙，足蹬尖头靴，脚踏卧牛台座，造型威武。

三彩井

Three-color Well

唐（618—907）
樊家遗址朱满墓出土
济南市考古研究院藏
高 6.02 厘米，口径最宽 7.83 厘米

四棱台状，中空，方形井口，井口上有仿木结构井框。

三彩臼

Three-color Mortar

唐（618—907）
樊家遗址朱满墓出土
济南市考古研究院藏
高 6 厘米，宽 5.1 厘米，长 17 厘米

长方形底座，中部有两立柱，立柱连接长杆，杆一端置杵。

三彩磨

Three-color Millstone

唐（618—907）
樊家遗址朱满墓出土
济南市考古研究院藏
高 3.8 厘米，口径 9.1 厘米

由磨盘和底座组成，磨盘顶部凸起一周挡粮圈。

三彩侍从俑

Three-color Attendant Figurine

唐（618—907）
樊家遗址朱满墓出土
济南市考古研究院藏
左：高 21.1 厘米，宽 5.3 厘米，厚 5.4 厘米
中：高 21.6 厘米，宽 5.6 厘米，厚 5.5 厘米
右：高 22.2 厘米，宽 5.8 厘米，厚 5.7 厘米

头戴幞（fú）头，身着圆领右衽窄袖长袍，双手交拱于胸前，手隐于袖中。

三彩女俑

Three-color Female Figurine

唐（618—907）
樊家遗址朱满墓出土
济南市考古研究院藏
高 23.4 厘米，宽 6.1 厘米

头梳双髻，上身穿白色窄袖襦衫，外套白色半臂，肩披绿色披帛，下穿黄色曳地长裙，露出圆头鞋。

三彩羊

Three-color Sheep

唐（618—907）
樊家遗址朱满墓出土
济南市考古研究院藏
上：高 13.5 厘米，宽 7.4 厘米，长 19.5 厘米
下：高 12.1 厘米，宽 6.1 厘米，长 19.5 厘米

羊头微抬，卷角，瞪眼，短尾下垂，体肥壮，立于底座上。

三彩鸡

Three-color Duck

唐（618—907）
樊家遗址朱满墓出土
济南市考古研究院藏
高 16 厘米，宽 5.8 厘米，长 12.8 厘米

尖喙小冠，鸡头挺拔，长尾下垂，通体随机施黄白绿三彩。

三彩猪

Three-color Pig

唐（618—907）
樊家遗址朱满墓出土
济南市考古研究院藏
高 12.2 厘米，宽 6.8 厘米，长 18.6 厘米

长嘴紧闭，鬃毛高耸，体肥壮，短尾贴附于左臀处，短腿，立于底座上。

金银平脱宝相纹葵花铜镜

Sunflower-shaped Bronze Mirror with Gold-Silver Pingtuo Inlay and Composite Floral Pattern

唐（618—907）
济南市中心医院唐代墓出土
济南市博物馆藏
直径 19 厘米

铜镜为六瓣形，圆钮，钮周围饰金片六出重瓣纹，每瓣为三重。其外为六个银片心形纹中套金片宝相纹，心形纹之间缀金片瓣纹。窄平素镜缘。此镜唐代墓葬出土，墓主人唐右威卫左中侯项承晖系“故赠使持节临淮诸军事临淮太守之次子，贵妃之令弟，公主之季舅”。

青釉鸡首魁

Celadon-glazed Chicken-Head Ewer (Kui)

北朝（439—581）
济南市趵伞塔附近出土
济南市博物馆藏
口径 27 厘米，腹围 75.8 厘米

口沿外饰鸡首低垂，鸡冠耸立，鸡眼圆睁，鸡嘴似钩，似欲觅食。汉代《韩诗外传》中将鸡称为具有文、武、勇、仁、信五德的“德禽”，两晋以鸡为装饰的瓷器盛行一时，并延续后朝。

三彩三足炉

Three-color Three-legged Furnace

唐（618—907）
济南铁路局技校院内出土
济南市博物馆藏
口径 11.6 厘米，腹围 55.4 厘米

贴塑别致，整器色彩鲜艳明亮，肩部堆贴 4 朵花卉。

府城之兴

济南自北宋时期升为府城，此后其政治、经济和文化地位日益凸显，建城活动持续进行。金元时期，济南成为山东地区的监察中心，政治地位进一步提高。明清时期，济南府成为山东省会，济南的城市建设得到全面提升，城墙、护城河等防御设施得到完善和加强；工商兴旺、文风昌盛，济南成为中国北方的重要中心城市。

济南府城（宋金元时期）

北宋，齐州先后属京东路和京东东路。政和六年（1116），齐州升为济南府，辖历城、章丘、长清等 5 县。建炎二年（1128）后，济南被金朝所据，仍为济南府，属山东东路。其间，曾一度为刘齐辖境。元初，济南改为济南路。至元二年（1265），辖棣州、滨州 2 州及历城、章丘、济阳、商河等 11 县。金元时期，济南先后为金山东东西路提刑司、元山东东西道肃政廉访司治所，是山东地区的监察中心。宋、金、元时期，济南成为一座人口密集、工商业繁荣的城市。

The Prosperity of Fucheng

Jinan was elevated to a prefecture city during the Northern Song Dynasty, and since then its political, economic, and cultural status became increasingly prominent, with continuois city building activities. During the Jin and Yuan dynasties, Jinan became the monitoring center of Shandong region, and its political status was further elevated. During the Ming and Qing dynasties, Jinan Prefecture became the capital of Shandong Province, and the urban construction of Jinan was comprehensively improved. Defense facilities such as city walls and moats were improved and strengthened; Jinan became an important central city in northern China due to its prosperous industry and commerce, as well as flourishing literary style.

Prefecture City of Jinan (Song, Jin and Yuan Dynasties)

In the Northern Song Dynasty, Qi Prefecture successively was divided into Jingdong Road and Jingdong East Road. In the sixth year of Zhenghe (1116), Qi Prefecture was elevated to Jinan Prefecture, which governed five counties including Licheng, Zhangqiu, and Changqing. After the second year of Jianyan (1128), Jinan was occupied by the Jin Dynasty and remained under the jurisdiction of Jinan Prefecture, divided into Shandong East Road. During that time, it was once under the jurisdiction of Liu Qi. At the beginning of the Yuan Dynasty, it was changed to Jinan Road. By the second year of the Yuan Dynasty (1265), it had jurisdiction over 2 prefectures of Dizhou and Binzhou, as well as 11 counties such as Licheng, Zhangqiu, Jiyang and Shanghe. During the Jin-Yuan Period, Jinan served as the administrative seat of the Shandong East-West Circuit Surveillance Office of Jin Dynasty and later Shandong East-West Circuit Disciplinary Inspectorate of the Yuan Dynasty, functioning as the judicial and monitoring center of the Shandong region. During the Song, Jin, and Yuan dynasties, Jinan became a densely populated city with thriving industry and commerce.

宋代府制

宋代行政区划实行州（府、军、监）、县二级制，同时在地方设置路，作为直辖于中央并高于府、州、军、监的一级监察区。

宋代在军事或经济上比较重要的城市设府加以统治，府下辖多个县。北宋最多时设 34 府，包括 4 处京府、30 处普通府。4 处京府为东京开封府（今河南开封）、西京河南府（今河南洛阳）、南京应天府（今河南商丘）、北京大名府（今河北大名）。南宋共设 42 府。

The Prefectural System of the Song Dynasty

The administrative divisions of the Song Dynasty implemented a two-level system of states (prefectures, armies and supervisors) and counties, and at the same time, roads were set up at the local level as a first level supervisory area directly under the central government and higher than the prefectures, states, armies and supervisors.

In the Song Dynasty, important cities in terms of military or economy were governed by prefectures, which had multiple counties under their jurisdiction. At its peak in the Northern Song Dynasty, there were a total of 34 prefectures, including 4 capital prefectures and 30 ordinary prefectures. The four capital prefectures were Dongjing Kaifeng Prefecture (now Kaifeng, Henan), Xijing Henan Prefecture (now Luoyang, Henan), Nanjing Yingtian Prefecture (now Shangqiu, Henan), and Beijing Daming Prefecture (now Daming, Hebei). A total of 42 prefectures were established in the Southern Song Dynasty.

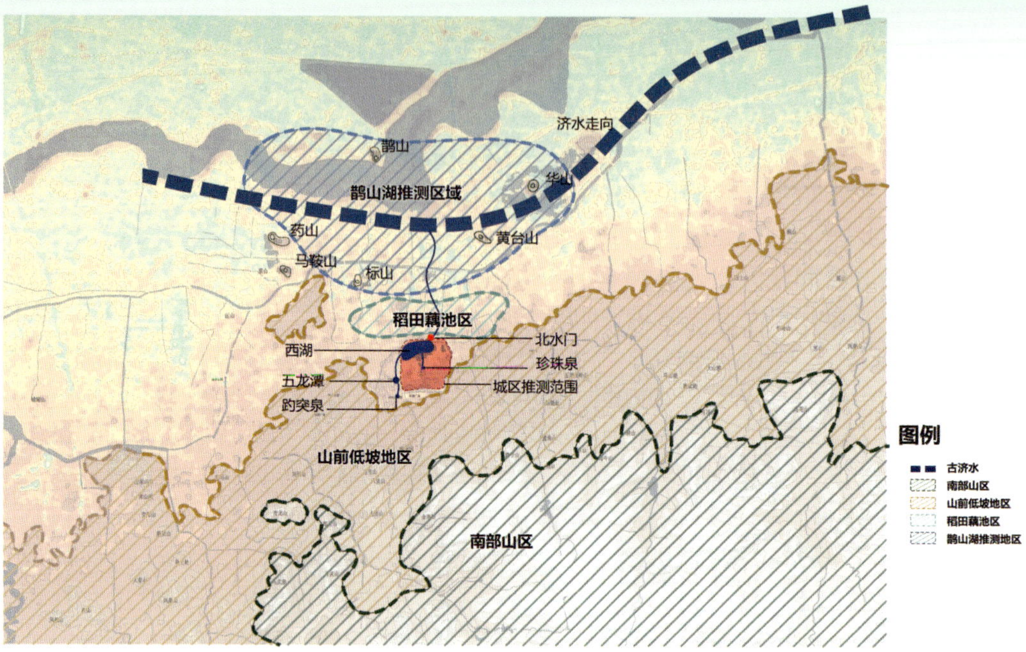

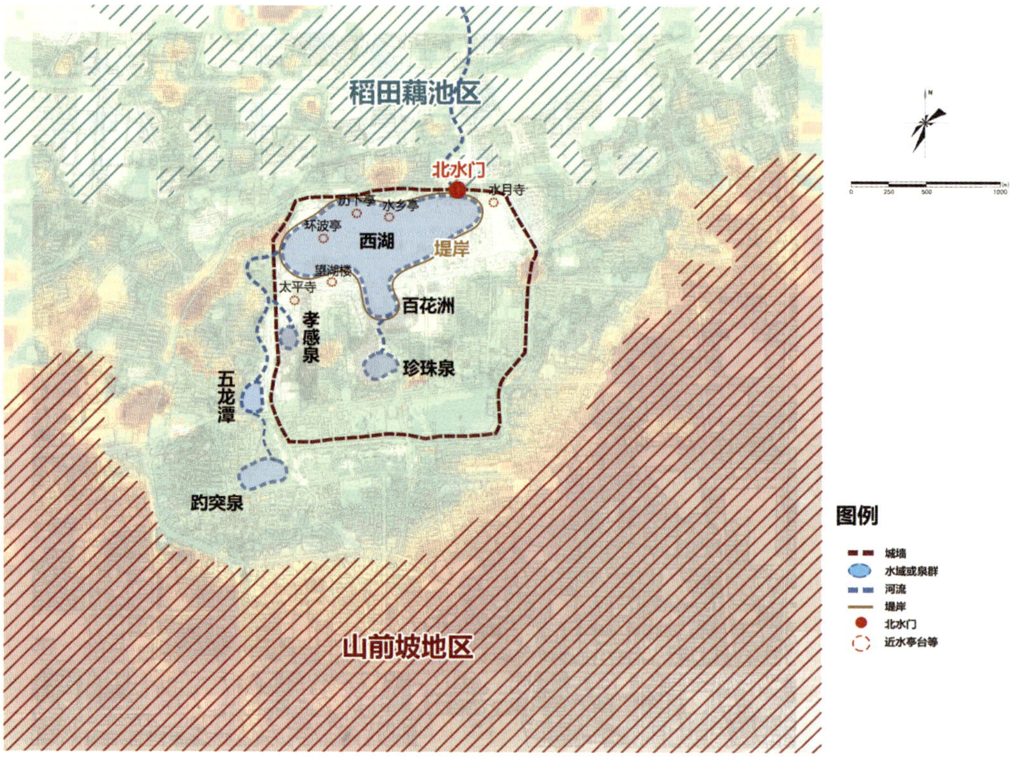

宋代的城市和水环境

Urban and Water Environment in the Song Dynasty

曾巩通过兴建各种水利设施，逐步对泉水环境进行人工控制和改造，包括在北城修筑堤堰、疏浚河道、开挖新渠、修建水闸等，取得显著成果。北水门因时调节泄水流量，使大明湖成为天然水库，既从根本上解决了水患，又有利于城北平原的灌溉。

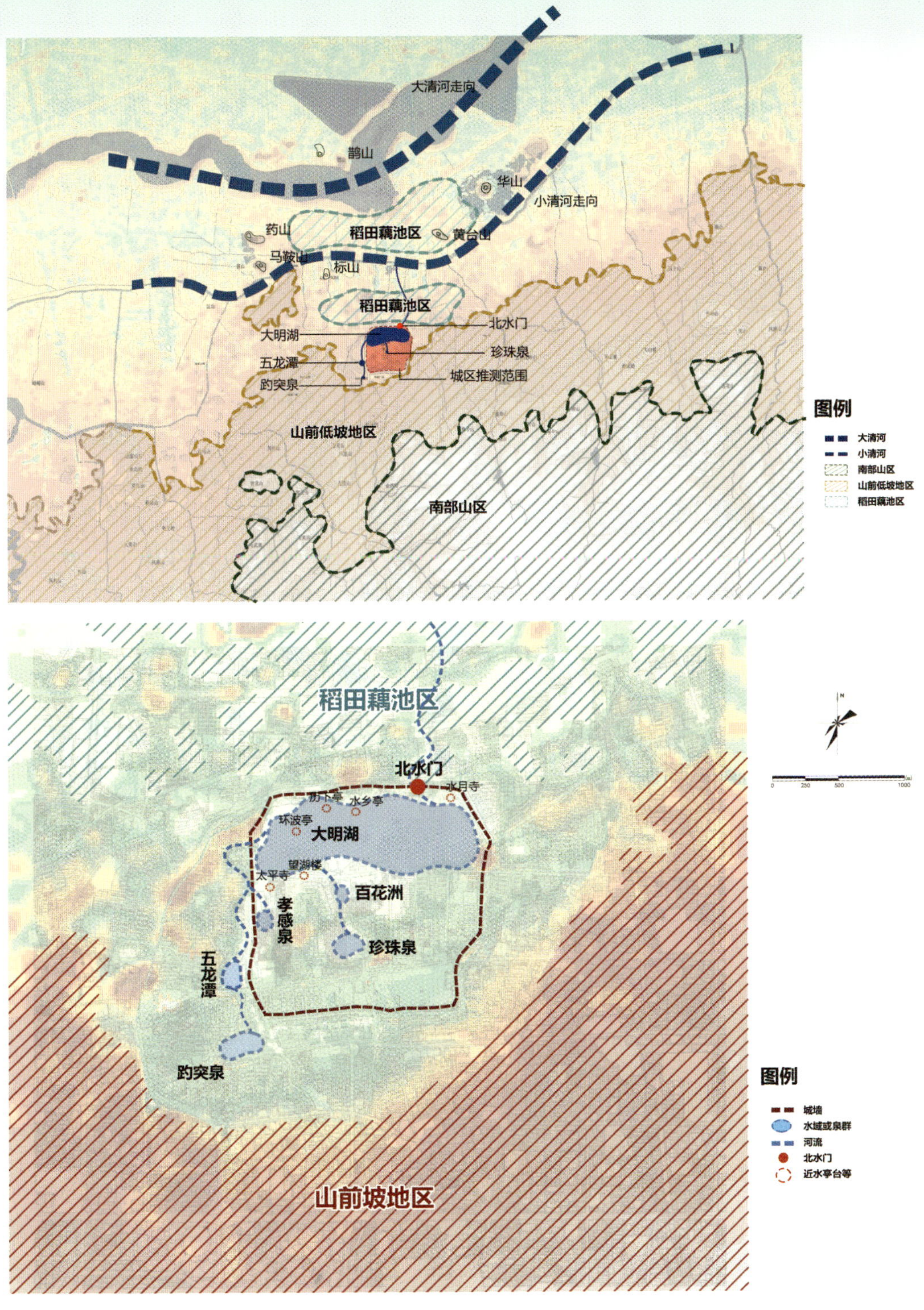

金元时期的城市和水环境

Urban and Water Environment During the Jin and Yuan Dynasties

金元时期，小清河的修建使济南北郊沼泽地区的积水得以宣泄，给北郊带来更多的田地。开凿小清河的大部分工程是疏浚挖掘旧有河道，同时也为裁弯取直、补充水源而挖掘了一些新河段，利用济南地区充沛的泉水保证河运水位，来往货船可由莱州湾驶至济南，便利了济南地区与山东半岛的物资交流，特别是食盐运输。

崇墉百雉——趵突泉北路 6 号古城墙遗址

古城西部的趵突泉北路 6 号遗址发现了宋代至明清时期的古城墙。古城墙整体呈东北—西南走向。城墙内侧凸出部分可能为“铺舍”（城垣顶上驻军值班房）、“旗台”（通过旗帜传递信号的场所）或“敌台”（用于防御或眺望敌人的楼台）之类的建筑物。残存墙体最高为 2.48 米，墙基最高为 2 米。古城墙纵截面呈上窄下宽的梯形，其内外两侧为砖石和石灰砌筑，中间为夯土版筑。

High and Strong City Wall: Historic Site of the Ancient City Wall on No. 6 Baotuquan North Road

At the Historic Site of the Ancient City Wall on No.6 Baotuquan North Road in the western part of the ancient city, ancient city walls from the Song Dynasty to the Ming and Qing dynasties were discovered. The overall direction of the ancient city wall is northeast-southwest. The protruding part on the inner side of the city wall may be a building called “Pu She” (a garrison duty room on the top of the city wall), “Qi Tai” (a place where signals were transmitted through flags), or “Di Tai” (a tower used for defense or observation of the enemy). The highest remaining wall is 2.48m, and the maximum height of the wall foundation is 2m. The longitudinal section of the ancient city wall is a trapezoid with a narrow upper part and a wide lower part. Its inner and outer sides are made of bricks, stones, and lime, and the middle is made of rammed earth plates.

趵突泉北路 6 号古城墙遗址
The Historic Site of the Ancient City Wall on No. 6 Baotuquan North Road

修筑年代

根据城墙夯土的叠压关系及包含物判断，这段城墙修建年代的上限应为宋代。在用于修筑城墙的条石中发现元末墓碑，结合《历乘》关于“济南府城墙于明洪武四年（1371）始内外包以砖石”的记载，城墙外侧的砖石砌筑应为明初开始修建。

防水措施

济南在古代就是水资源丰富的城市，排水、防水是筑城所要解决的重要问题。在城墙外壁下端有鹅卵石铺成的散水，散水呈斜坡状，越向城外越低，而在城墙内侧发现了一个围绕城墙的大型池塘。

《历城县志》载：“砖陂城以石为趾，砖为肤，土为骨，阔凡五丈。近内者多陂，上覆以甓，各三尺许……”即用石料作为城墙基础，用砖包土的做法建造墙体。位于城内的墙体靠近多水区域，通常会覆盖更厚的砖，以达到“防水潦之浸灌”的目的。

Age of Construction

Judging from the stacking relationship and contents of the rammed earth of the city wall, the upper limit of the construction period of this section of the city wall should be the Song Dynasty. Tombstones from the end of the Yuan Dynasty was discovered among the stones used to build the city wall. Combined with the record in the *Li Sheng* that “the walls of Jinan Prefecture were surrounded by bricks and stones from the fourth year of the Hongwu of the Ming Dynasty (1371)”, the masonry with bricks and stones on the outer side of the city wall should have started in the early Ming Dynasty.

Waterproofing Measures

Jinan was a city rich in water resources in ancient times, and drainage and waterproofing were important issues to be addressed in the construction of the city. At the lower end of the outer wall of the city wall, there is apron paved with cobblestone. The apron is in a sloping shape, decreasing towards the outside of the city, and a large pond surrounding the city wall has been discovered on the inner side.

As recorded in *Annals of Licheng County*, “Zhanbei City is made of stone as the toe, brick as the skin, and soil as the bone, and it is five feet wide. Near the inner side, multiple stone-reinforced embankments were constructed, each topped with brick cladding of approximately three chi (about 1m) in thickness...” The walls located in the city are close to areas with high water content and are generally covered with thicker bricks to achieve the purpose of “prevention of flooding by waterlogging”.

趵突泉北路 6 号古城墙遗址考古发掘现场
The Archaeological Excavation Site of the Ancient City Wall at No.6, Baotuquan North Road

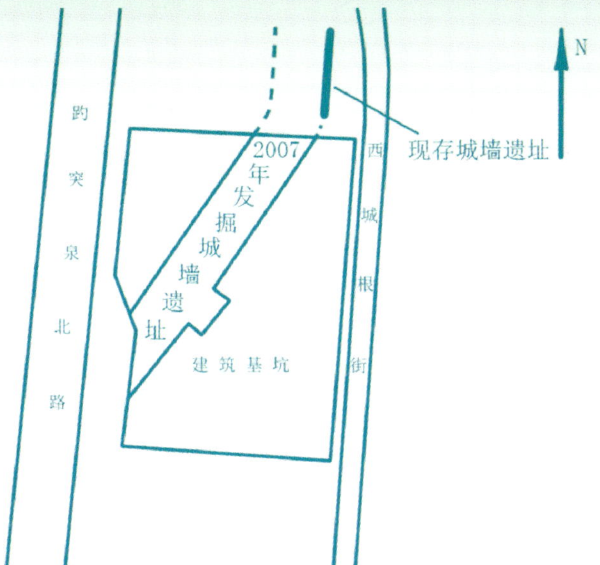

趵突泉北路6号古城墙遗址平面图
Floor Plan of the Historic Site of the Ancient City Wall on No. 6 Baotuquan North Road

总长约80米；凸出部分宽15.9米，其北侧宽12.7米、南侧宽12米

城墙上的夯窝
Rammed Pits on the City Wall

城墙横断面
Cross Section of the City Wall

石条砌筑的外侧城墙墙体
Building Wall of the Masonry City Wall on the Outer Side

石条砌筑的内侧城墙墙体
Building Wall of the Masonry City Wall on the Inner Side

市井生活

旧军门巷遗址

古城西南部的旧军门巷一带，自商代至明清时期均有人类生活。旧军门巷遗址发现反映宋代到明清市井生活的房址 8 座、水井 2 眼、水窖 1 口，较完整器物 30 余件及大量陶、瓷片，极大地丰富了陶、瓷器的研究资料。

Traces of Daily Life at That Time

Historic Site of Former Junmen Lane

The area around Former Junmen Lane in the southwest of the ancient city was inhabited by humans from the Shang Dynasty to the Ming and Qing dynasties. At the historic site of Former Junmen Lane, 8 house sites from different periods, 2 water wells, a water cellar of Daily Life from the Song Dynasty to the Qing Dynasty, more than 30 relatively complete artifacts, as well as a large number of pottery and porcelain pieces, have been discovered, significantly enriching the research data on pottery and porcelain.

旧军门巷遗址考古发掘现场
Excavation site of Former Junmen Lane

发掘出的晚期井
A Well of the Late Period Excavated

探方内一眼古井
An Ancient Well in the Excavation Trench

瓷塑像

Porcelain Statue

宋（960—1276）
旧军门巷遗址出土
济南市考古研究院藏
高 18 厘米

头戴花冠，面容慈祥，双目微睁，平视前方，面庞圆润，双唇紧闭。

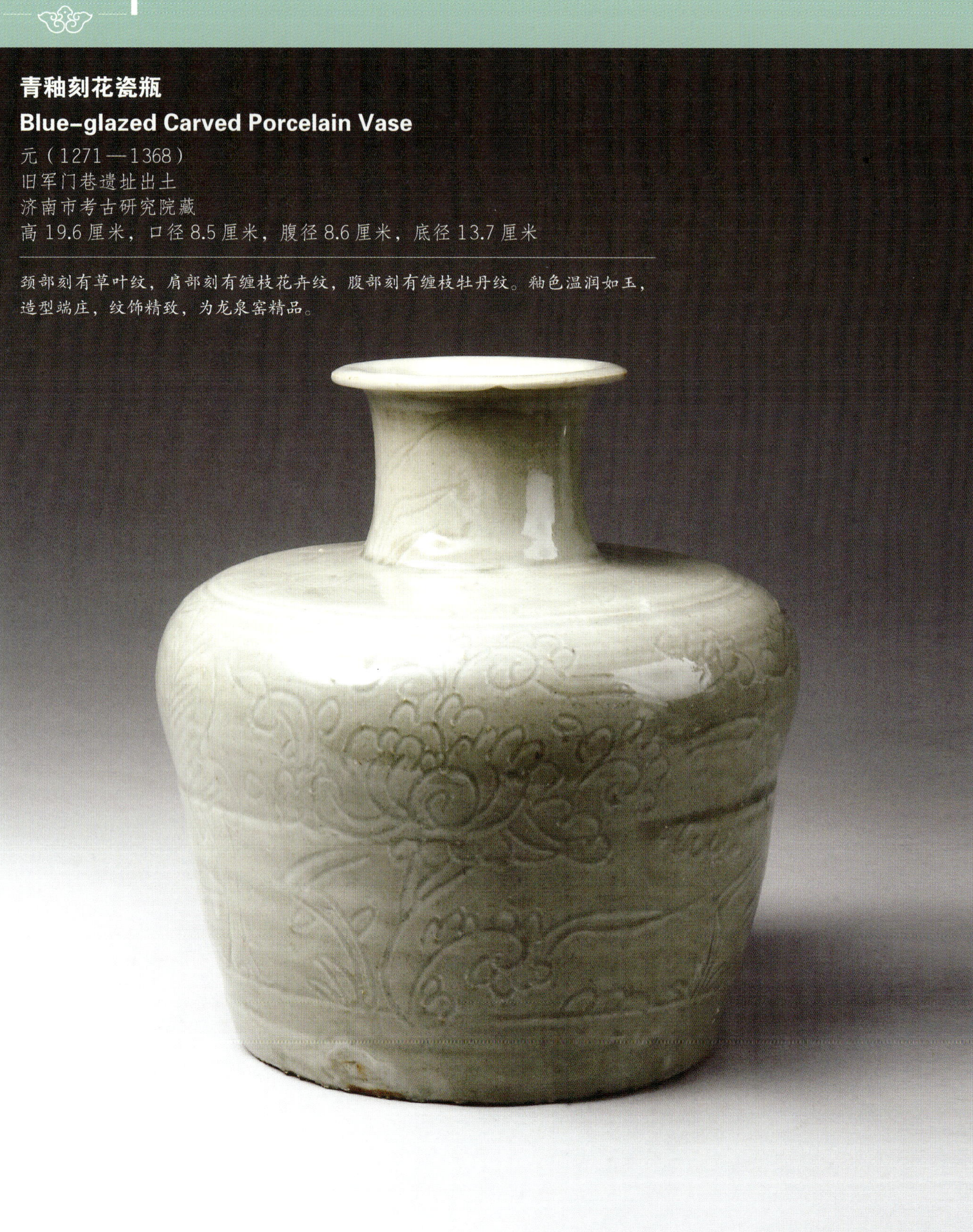

青釉刻花瓷瓶

Blue-glazed Carved Porcelain Vase

元（1271—1368）
旧军门巷遗址出土
济南市考古研究院藏
高 19.6 厘米，口径 8.5 厘米，腹径 8.6 厘米，底径 13.7 厘米

颈部刻有草叶纹，肩部刻有缠枝花卉纹，腹部刻有缠枝牡丹纹。釉色温润如玉，造型端庄，纹饰精致，为龙泉窑精品。

白釉瓷罐
White-glazed Porcelain Jar

元（1271—1368）
高都司巷遗址出土
济南市考古研究院藏
高 5.7 厘米，口径 8.7 厘米，底径 5.2 厘米，腹径 9.6 厘米

圆唇，敞口，垂鼓腹，圈足。器身施白釉至下腹部，露灰白色夹砂胎，器物内施透明釉，釉内有气泡。

白釉瓷碗
White-glazed Porcelain Jar

宋（960—1276）
旧军门巷遗址出土
济南市考古研究院藏
高 8.2 厘米，口径 18.63 厘米，底径 6 厘米

圆唇，敞口，弧腹，圈足。胎体较薄。内外通体施白色釉。

钧釉瓷碗

Jun-glaze Porcelain Bowl

元（1271—1368）
旧军门巷遗址出土
济南市考古研究院藏
高 7.8 厘米，口径 18.6 厘米，底径 6 厘米

敞口，尖唇，深弧腹，小圈足。天蓝釉，釉层厚，质感强，外釉呈流淌状及足。

十亩园墓群

十亩园墓群位于济南市历下区十亩园七家村，济南古城墙（解放阁）以东约 150 米。1998 年的发掘共清理宋代墓葬 57 座、灰坑 17 座，出土大量珍贵文物。发掘如此规模的宋代墓群当时在山东尚属首次，在全国范围内也较为罕见。清理古墓群时，发现在该墓群的宋代堆积层下还有隋唐文化层，含有隋唐时期的陶瓷片，这说明在隋唐时期这一区域已有先民居住，这一发现对于研究济南城市变迁具有重要意义。

Shimuyuan Tombs

The Shimuyuan Tombs are located in Qijia Village, Shimuyuan, Lixia District, Jinan City, approximately 150m east of the ancient city wall (Jiefang Ge) of Jinan. In 1998, a total of 57 tombs and 17 ash pits from the Song Dynasty were excavated, and a large number of precious cultural relics were unearthed. Discovery of such large-scale tombs of the Song Dynasty was the first of its kind in Shandong and relatively rare nationwide. When cleaning up the ancient tombs, it was discovered that the tombs also had a cultural layer of the Sui and Tang dynasties beneath the accumulation layer of the Song Dynasty, containing ceramic fragments from the Sui and Tang dynasties. It indicates that the area was inhabited by ancients during the Sui and Tang dynasties, and the discovery is of great significance for the study of urban changes in Jinan.

白釉瓷香薰

White-glazed Porcelain Aroma Burner

宋（960—1276）
十亩园宋代墓群出土
济南市考古研究院藏
高 7.9 厘米，口径 7.08 厘米，底径 5.7 厘米

整器内外均施白釉。该薰炉形似盖豆，子母口盖，炉盖镂空，为半圆形，顶部中央镂刻一圆形孔，周围一圈雕刻有 10 个等大旋涡状小孔，形似一盛放花朵，再外镂刻两圈菱形孔及一圈三角形孔，疏密相间，错落有致。

青釉“李宅”瓷梅瓶

Green-glazed “Li Zhai” porcelain Prunus Vase

宋（960—1276）
按察司街遗址出土
济南市考古研究院藏
高 39.4 厘米，口径 5.2 厘米，底径 8 厘米，腹径 20.2 厘米

瓶小口，折沿，短颈，丰肩，肩以下渐收，圈足。瓶身上半部分施青釉，下半部分施酱釉。足底露胎。由肩部向下斜向墨书“李宅”二字。整器优美修长，秀丽挺拔，形似美人。

白釉黑花瓷碗

White glazed Porcelain Bowl with Black Flowers

金（1115—1234）
按察司街遗址出土
济南市考古研究院藏
高 4.9 厘米，口径 12.4 厘米，底径 5 厘米

侈口，弧腹，圈足。器内外采用白地黑花的装饰手法，口沿内饰两周弦纹夹草叶纹，碗内饰一周弦纹，中心与四周均匀分布五朵花卉纹饰。

龙泉窑龙纹玉壶春瓶

Longquan Kiln Celadon Yuhuchun Vase with Dragon Pattern

元（1271—1368）
济南消防汽车装配厂出土
济南市博物馆藏
高 26 厘米，口径 6.8 厘米，底径 7.4 厘米，腹围 44 厘米

腹部凸印一飞龙戏火珠。龙双目圆睁，张口，发须飞扬。龙纹保持有宋代的特征。

釉里红折枝花纹玉壶春瓶

Under-glaze Yuhuchun Vase with Red "Pruned-Branch" Patterns

元（1271—1368）
大观园出土
济南市博物馆藏
口径 7 厘米，底径 6.9 厘米，腹围 38.5 厘米

整器淡青色釉地，釉色润亮，釉里红纹饰稍显发暗。此器造型美观，釉面细润，纹饰简洁雅致，画工娴熟，是典型的元代器物。

青白釉瓷炉

Blue and white-glazed Porcelain Furnace

宋（960—1276）
按察司街遗址出土
济南市考古研究院藏
高 8.12 厘米，口径 7.9（最宽处）厘米，底径 5.6 厘米

宽平沿，灯身深腹碗式，莲花座。通体施青白釉，影青色，器质细腻，釉色晶莹，白中泛青，胎体洁白致密，做工精美。

青釉瓷钵

Celadon-glazed Porcelain Furnace

元（1271—1368）
按察司街遗址出土
济南市考古研究院藏
口径 10.9 厘米，底径 4.8 厘米，高 5.4 厘米

敛口，圆唇，鼓腹下渐收。施青釉，足部素烧。开片纹。

白釉瓷俑

White-glazed Porcelain Figurine

宋（960—1276）
按察司街遗址出土
济南市考古研究院藏
宽 2 厘米，高 5.7 厘米

人物造型为一中老年男性，头顶中部梳一短髻，余发披散至肩，双目微睁，神态自若，身着宽袍，衣襟松垂，衣裾及地，仅露鞋尖。衣褶线条简练，双手搭于胸前，左手持一圆环状物。

最早的广告商标

北宋时期，济南地处水陆交通要道，商品经济、市镇经济繁荣，店铺依街布设，逐渐发展成为商品贸易的重要集散地。“济南刘家功夫针铺”商标铜版的出现，表明当时的手工业已开始利用商标作为保护和推销产品的手段，也反映了济南城市商品经济的发展，这是迄今为止我国发现的最早的广告商标，也是目前已知世界上最早出现商标的广告。

The Earliest Advertising Logo

During the Northern Song Dynasty, Jinan was located on an important land and water transportation route. The commodity economy and the town economy were prosperous. Shops were arranged along the streets, and it gradually developed into an important distribution center for commodity trade. The appearance of the copper plate of the trademark of "Jinan Liujia Kung Fu Needle Shop" shows that the handicraft industry at that time had begun to use trademarks as a means of protecting and promoting products, and it also reflects the development of Jinan's urban commodity economy. This is the earliest advertising logo discovered in China so far, and it is also the earliest known trademark advertisement in the world.

“济南刘家功夫针铺”广告青铜版（中国国家博物馆藏）
The Bronze Advertising Board of Jinan Liujia Kung Fu Needle Shop(Collected by National Museum of China)

金银窖藏——卫巷遗址

古城南部的卫巷遗址一窖穴中，出土了一罐北宋时期的金银器，其中金器 10 件、银器 25 件。金银器窖藏成因或许与北宋晚期的战乱有关。卫巷遗址内的遗存，其年代涉及两周至清代，文化序列延续清晰，以宋、金、元遗存最为丰富。

Gold and Silver Cellar: Weixiang Historic Site

In a cellar of Weixiang Historic Site in the southern part of the ancient city, a jar of gold and silver artifacts from the Northern Song Dynasty was unearthed, including 10 gold artifacts and 25 silver artifacts. The cause for the cellar storage of gold and silver artifacts may be related to the wars in the late Northern Song Dynasty. The remains of Weixiang Historic Site date back to the Zhou Dynasty to the Qing Dynasty, and the cultural sequence continues clearly, with the Song, Jin and Yuan dynasties being the most abundant.

卫巷名称的由来

卫巷，因明代位于济南卫西面而得名。“卫”是我国明代军队的编制名称，当时于战略要地设“卫”，每卫约 5600 人。

Origin of the Name of Weixiang

Weixiang was named its position west of the Jinan Wei Garrison during the Ming Dynasty. “Wei Garrison” is the formation of the army of the Ming Dynasty in China. At that time, “Wei Garrisons” was set up in strategic locations, with each Wei Garrison consisting of about 5,600 people.

卫巷遗址考古发掘探方
Exploration Trench for Archaeology at Weixiang Site

黑釉双系瓷罐

Black-glazed Double-series Porcelain Pot

宋（960—1276）
卫巷遗址出土
济南市考古研究院藏
高 18.3 厘米，口径 12.9 厘米，足径 9.2 厘米，腹径 19.1 厘米

直口略敛，圆肩，肩上双桥形耳，深鼓腹略垂，圈足；施黑釉，釉色光亮，釉不及底，尖唇部成黄色。

化生童子金耳坠

Gold Earrings with Buddhist "Transformation-Born Child" Pattern

宋（960—1276）
卫巷遗址出土
济南市考古研究院藏
左：长 2.88 厘米，宽 1.3 厘米
右：长 2.9 厘米，宽 1.37 厘米

一对。使用锤揲（dié）、錾（zàn）刻、镂空、掐丝等工艺打造为化生童子形象。童子头戴花冠，脸庞丰润，眉眼清晰，手持莲花，繁杂璎珞缠身，立于莲花座上。

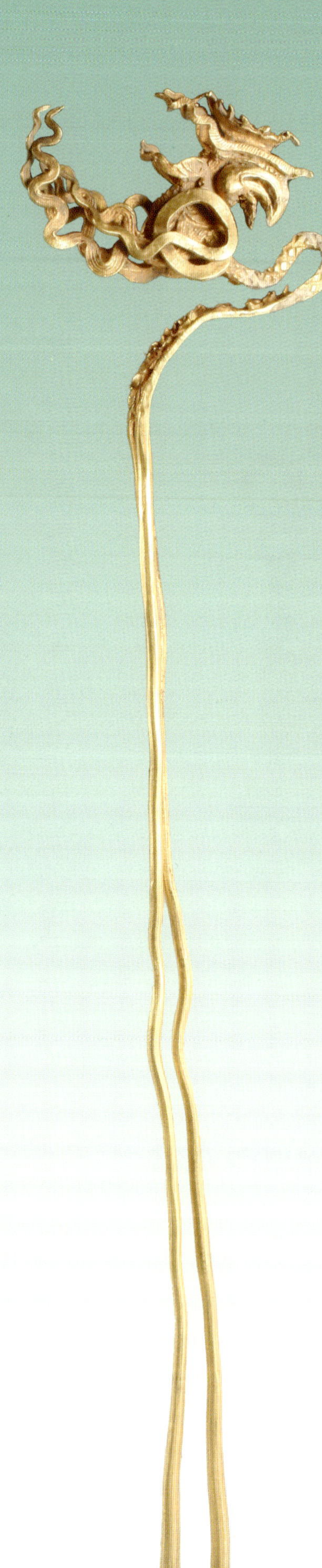

凤头金钗

Phoenix-headed Gold Hairpin

宋（960—1276）
卫巷遗址出土
济南市考古研究院藏
长 22.6 厘米，宽 4.4 厘米

簪首透雕，作凤鸟状，凤翅高高翘起，周围绫缎缠绕，作振翅高飞之势。鸟冠处有镶嵌宝石的凹槽，出土时宝石已不存。

刻花金钗

Engraved Gold Hairpin

宋（960—1276）
卫巷遗址出土
济南市考古研究院藏
长 20.9 厘米，宽 1.1 厘米

钗梁顶端及钗脚处均錾刻有“吴二郎造”款识。

素面金钗

Plain Gold Hairpin

宋（960—1276）
卫巷遗址出土
济南市考古研究院藏
长 20 厘米

钗梁光素无纹饰，顶端及钗脚处均錾刻有“足金”款识。

素面金钗

Plain Gold Hairpin

宋（960—1276）
卫巷遗址出土
济南市考古研究院藏
长 14.5 厘米

钗梁光素无纹饰。

素面金环

Plain Gold Ring

宋（960—1276）
卫巷遗址出土
济南市考古研究院藏
左：直径 2.57 厘米，孔径 1.4 厘米，厚 0.16 厘米
右：直径 2.7 厘米，孔径 1.4 厘米，厚 0.14 厘米

金环制作方式为金片锤揲成扁平璧状，光素无纹饰。

树叶形金耳坠

Leaf-shaped Gold Earrings

宋（960—1276）
卫巷遗址出土
济南市考古研究院藏
左：长 3.2 厘米，宽 2 厘米，高 3 厘米
右：长 3.4 厘米，宽 2 厘米，高 3 厘米

一对。装饰部分锤揲、錾刻为高浮雕的牡丹花叶。

古寺地宫——县西巷遗址

县西巷遗址位于古城中部，遗址内发现汉唐、宋元、明清时期的水井、窖穴、炉灶、道路、地宫等遗迹及数百件瓷器、陶器、建筑构件等遗物。其中，县西巷宋代砖雕地宫是目前国内发现的同时期地宫中雕刻最为精美的，地宫中出土石碑《开元寺修杂宝经藏地宫记》，记载了当时地宫的修建情况。

The Underground Palace of the Ancient Temple: The Historic Site of Xianxi Lane

The Xianxi Lane Historic Site is located in the central part of the ancient city. Within the site, relics such as water wells, cellars, stoves, roads and underground palaces from the Han, Tang, Song, Yuan, Ming and Qing dynasties have been discovered, as well as hundreds of porcelain, pottery, architectural components and other artifacts. Specifically, the brick-carved underground palace in Xianxi Lane of the Song Dynasty is currently the most exquisitely carved earth palace of its era in China during the same period. A stone tablet titled *Record of the Underground Palace of the Miscellaneous Treasures Collection in Kaiyuan Temple* was unearthed in the underground palace, documenting the construction construction of the palace at that time.

县西巷名称的由来

县西巷因位于明清历城县署西侧而得名。

Origin of the Name of Xianxi Lane

Xianxi Lane was named after its location on the west side of the Licheng County Office in the Ming and Qing Dynasties.

县西巷宋代砖雕地宫发掘现场
The Excavation Site of the Brick-carved Underground Palace of the Song Dynasty in Xianxi Lane

地宫残部
Remnants of the Underground Palace

地宫砖雕图案
Underground Palace Brick-carved Pattern

地宫砖雕图案
Underground Palace Brick-carved Pattern

《开元寺修杂宝经藏地宫记》拓片

开元寺的历史可追溯至隋唐，最初的寺址在济南古城范围内。《开元寺修杂宝经藏地宫记》记载了地宫的建造始末，也是开元寺最初寺址的有力证明。明朝初年，位于古城中心的开元寺改建成济南府署（即今县西巷东邻的省政协）。济南城内的开元寺遂告终结，寺内僧众迁到佛慧山的佛慧寺中，佛慧寺遂改称开元寺。自此以后，佛慧寺的香火被开元寺的佛光所代替。

Rubbing of *Record of the Underground Palace of the Miscellaneous Treasures Collection in Kaiyuan Temple*

The history of Kaiyuan Temple can be traced back to the Sui and Tang dynasties, with the original temple site located within the ancient city of Jinan. *Record of the Underground Palace of the Miscellaneous Treasures Collection in Kaiyuan Temple* documents the entire construction process of the underground palace and serves as a crucial archaeological evidence for the original site of Kaiyuan Temple. In the early years of the Ming Dynasty, Kaiyuan Temple located in the center of the ancient city was rebuilt into Jinan Prefecture Office (now the Provincial Political Consultative Conference adjacent to Xixiang in the county). The Kaiyuan Temple within Jinan city thus came to an end. Its monastic community relocated to Fohui Temple on Mount Fohui, which was subsequently renamed Kaiyuan Temple. Since then, the incense of Fohui Temple was replaced by the Buddha Light of Kaiyuan Temple.

县西巷祭坛和众石雕佛造像
The Altar of Xianxi Lane and the Statues of Stone-carved Buddhas

在县西巷地宫南侧发现了佛教举行宗教仪式的“坛”，这在全国已知的佛教造像埋藏坑中尚属首次。

出土唐代石雕力士像
Unearthed Stone Statue of of Vajrapani (Buddhist Guardian Warrior) from the Tang Dynasty

彩绘贴金圆雕菩萨头像

Painted gold-inlaid round carved Bodhisattva Head

北齐（550—577）
县西巷遗址出土
济南市考古研究院藏
长 20 厘米，宽 18 厘米，高 28 厘米

青石质，贴金彩绘。头戴精美花冠，面容丰满，双目细长，鼻残，唇线分明。额前发际中分，慈眉善目，神态安详。

佛教造像

佛教自东汉传入中国后，在南北朝至隋唐五代时期取得空前发展。由于当时寺院和僧侣享有特权，对经济社会发展产生不利影响，遂发生“三武一宗”灭佛事件。县西巷遗址出土雕刻精美、形态各异的北朝至宋代佛像 80 余尊，但多数残缺不全，这正是对当时社会运动的一种真实反映。

Buddhist Statues

After Buddhism was introduced to China during the Eastern Han Dynasty, it achieved unprecedented development from the Southern and Northern Dynasties to the Sui, Tang and Five dynasties period. Due to the privileges enjoyed by temples and monks at that time, which had a negative impact on economic and social development, the "Three Wus (Taiwu of Northern Wei, Wu of Northern Zhou, and Wuzong of Tang) and One Zong (Shizong of Later Zhou) Imperial Suppression of Buddhism" subsequently occurred. More than 80 exquisitely carved and diverse Buddha statues from the Northern Dynasties to the Song Dynasty were unearthed from the Xianxi Lane, but most of them are incomplete, which is a true reflection of the social movement at that time.

圆雕菩萨头像

Round-carved Bodhisattva Head

北齐（550—577）
舜井街遗址出土
济南市考古研究院藏
长 12 厘米，宽 14 厘米，高 30 厘米

石质。头戴花冠，发际平整，面带微笑，神态安详。

坐佛像

Seated Buddhist Statue

唐（618—907）
县西巷遗址出土
济南市考古研究院藏
宽 23.9 厘米，高 38 厘米，厚 19.5 厘米

肩部以上及右臂残失，左手抚膝，着袒右僧祇支，结跏趺坐于台上。

背屏式立姿佛像

A Standing Buddha Statue with a Back Screen

北朝（439—581）
县西巷遗址出土
济南市考古研究院藏
长 15 厘米，宽 7.6 厘米，高 29.8 厘米

背屏素面，佛像面部不清，着厚重佛衣，左手施无畏印，赤足立于平台上。

捧宝函弟子像

Portrait of A Disciple Holding a Treasure Box

唐（618—907）
县西巷遗址出土
济南市考古研究院藏
长 12 厘米，宽 11.5 厘米，高 35 厘米

头部残失，身形修长，双手捧一函于胸前，赤足而立。推测为阿难。

县西巷遗址出土佛像
Buddha Statues Unearthed From the Site of Xianxi Lane

立姿菩萨像

Standing Bodhisattva Statue

北齐（550—577）
县西巷遗址出土
济南市考古研究院藏
长 14.7 厘米，宽 7.5 厘米，高 34.2 厘米

头部、手臂、底座残失。体态修长挺拔，衣裙贴体，披帛飘逸，璎珞繁丽，在身前形成“X”形交叉，中间结为花叶形，是典型的北齐青州风格造像。

菩萨像

Bodhisattva Statue

唐（618—907）
县西巷遗址出土
济南市考古研究院藏
高 27.2 厘米，宽 8.4 厘米，厚 7.2 厘米

头部与左臂残失，袒上身，下身着贴体长裙。佩戴璎珞圈，披帛绕颈沿双臂自然下垂，赤足立于仰莲座上。

一铺三身像

A shrine with statues of the Three Buddha Bodies

宋（960—1276）
县西巷遗址出土
济南市考古研究院藏
长 32.5 厘米，宽 15 厘米，高 42 厘米

正面为一铺三身像，中间为结跏趺坐于莲台上的佛像，肉髻螺发，双手合十，头光呈桃形；两侧胁侍立于小莲台上。背面为唐代线刻毗沙门天王像，两侧面线刻菩萨像。该像推测为宋代利用唐代构件雕刻。

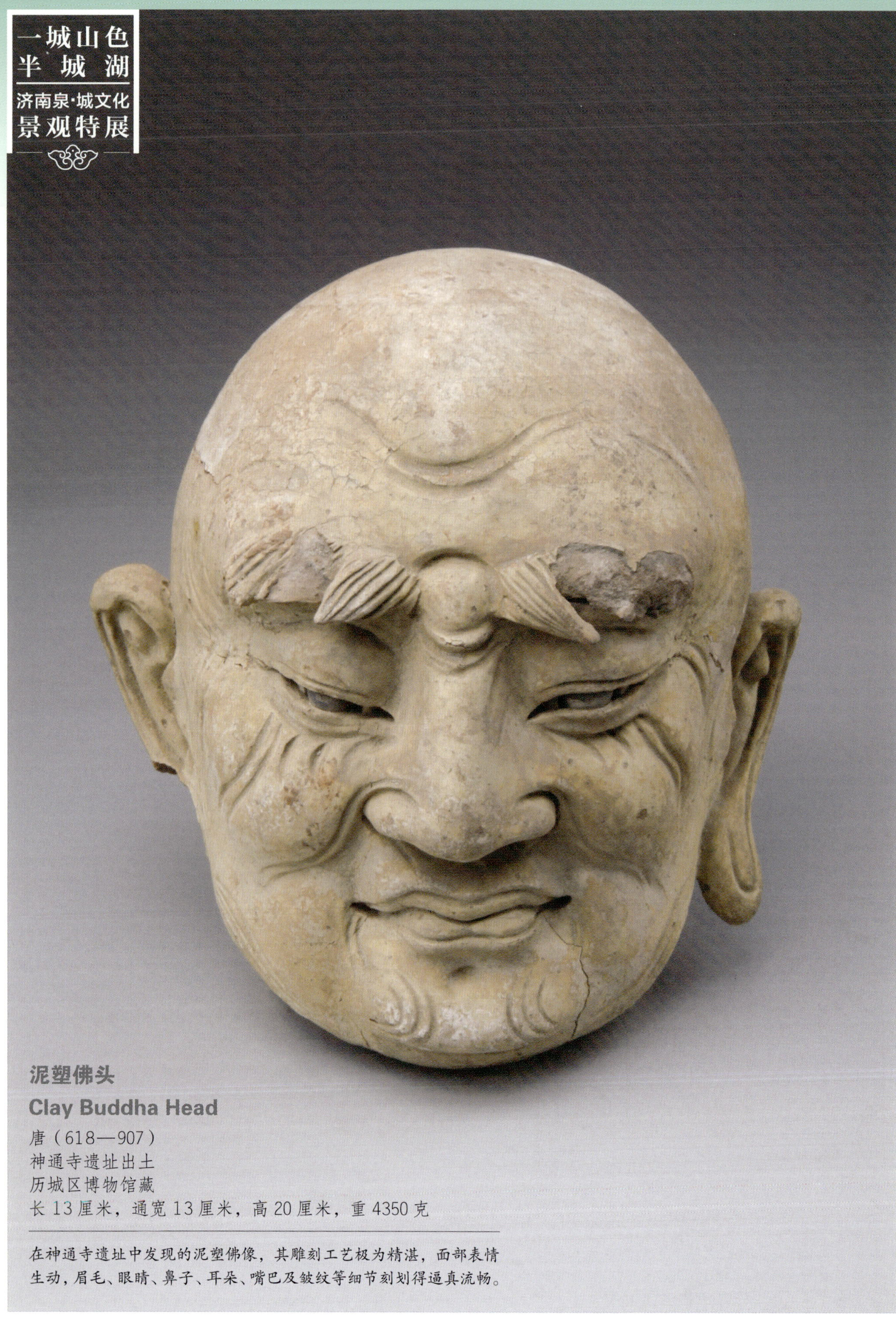

泥塑佛头

Clay Buddha Head

唐（618—907）
神通寺遗址出土
历城区博物馆藏
长 13 厘米，通宽 13 厘米，高 20 厘米，重 4350 克

在神通寺遗址中发现的泥塑佛像，其雕刻工艺极为精湛，面部表情生动，眉毛、眼睛、鼻子、耳朵、嘴巴及皱纹等细节刻划得逼真流畅。

四面佛像

Four-sided Buddha Statue

北齐（550—577）
舜井街出土
济南市博物馆藏
长 48.5 厘米，宽 42.5 厘米，高 62 厘米

造像略呈方柱型，四面都凿有佛龛（kān），龙首龛楣，其中三面佛龛内雕一佛二菩萨，一面佛龛内雕一佛二弟子。造像阴刻、浮雕、圆雕相互配合，层次分明，衣纹线条流畅，雕刻技艺精湛，显示出北朝晚期风格。

世侯气象——元济南王张荣家族墓地

元代济南王张荣家族墓地是中国迄今发现的级别最高、陵园附属物最多、一次性出土文字资料最丰富的元代墓地，位于历城区章灵丘村以北。经考古发掘，初步判断相关墓葬有 32 座，其中张荣墓是全国发现的规模最大、结构最复杂、壁画最丰富的元代墓葬，其前后双门楼、八墓室的结构在国内元代墓葬中为首次发现。

The Magnificence of Hereditary Nobility: The Family Cemetery of Zhang Rong, Prince of Jinan of the Yuan Dynasty

The tombs of Zhang Rong's Family, the Prince of Jinan during the Yuan Dynasty, is the highest level tomb discovered in China so far, with the the most cemetery appurtenances and the richest collection of written materials unearthed at one time. It is located north of Zhanglingqiu Village in Licheng District. According to preliminary archaeological excavations, there are 32 related tombs, among which Zhang Rong's tomb is the largest, most complex in structure, and richest in murals discovered in China during the Yuan Dynasty. Its front and rear double-gatehouse buildings and eight tomb chambers are the first discovery in the tombs of the Yuan Dynasty in China.

M83 张荣墓全景（上为西）
Panoramic View of Zhang Rong's Tomb M83 (West in the Top)

元济南王张荣

张荣（1181—1263），字世辉，济南历城人，金末率乡民起义，占据鲁北大片地区，成为金蒙之际山东三大世侯之一，与真定史氏、东平严氏、顺天张氏并称元代“四大诸侯”。1226年，张荣归顺蒙古，授金紫光禄大夫、山东行尚书省，兼兵马都元帅，知济南府事。1230年，朝廷集中诸侯议取汴梁，张荣请为开路先锋，元太宗喜之，赐衣三袭，位列诸侯之上。在征战过程中，张荣屡建奇功，并多次谏止屠杀。1236年，张荣将治所移至济南府。1260年，张荣被元世祖授予济南路万户，并封济南公。1263年，张荣以83岁的高龄去世，追封济南王，谥忠襄。

Zhang Rong, Prince of Jinan of the Yuan Dynasty

Zhang Rong (1181–1263), styled Shihui, was a native of Licheng, Jinan. During the late Jin Dynasty, he mobilized local militias to establish autonomous control over vast territories in northern Shandong, emerging as one of the "Three Great Hereditary Lords (Shihou) of Shandong" during the Jin-Mongol transition. He was later grouped with the Shi clan of Zhending, the Yan clan of Dongping, and the Zhang clan of Shuntian as the "Four Major Hereditary Princes" of the Yuan dynasty. In 1226, following his surrender to the Mongol Empire, Zhang Rong was granted three key appointments, including Grand Master of Imperial Entertainments with Golden Seal and Purple Ribbon, Secretariat Branch Administrator for Shandong, and Supreme Military Commander and Prefect of Jinan. In 1230, when the Mongol court convened a council of hereditary lords to strategize the capture of Bianliang, Zhang Rong petitioned to lead the vanguard. Emperor Taizong of Yuan was delighted by his initiative, awarding him three ceremonial robes (a Mongol imperial honor symbolizing trust), and elevating his rank above all other hereditary lords. During the course of the battles, Zhang Rong repeatedly made miraculous achievements and admonished the massacre for many times. In 1236, Zhang Rong moved his county town to Jinan Prefecture. In 1260, Zhang Rong was appointed as "Commander of Jinan Circuit" and enfeoffed as "Duke of Jinan" by Emperor Shizu of Yuan Dynasty. In 1263, Zhang Rong passed away at the age of 83 and was posthumously honored as the King of Jinan, with the posthumous title of Zhongxiang.

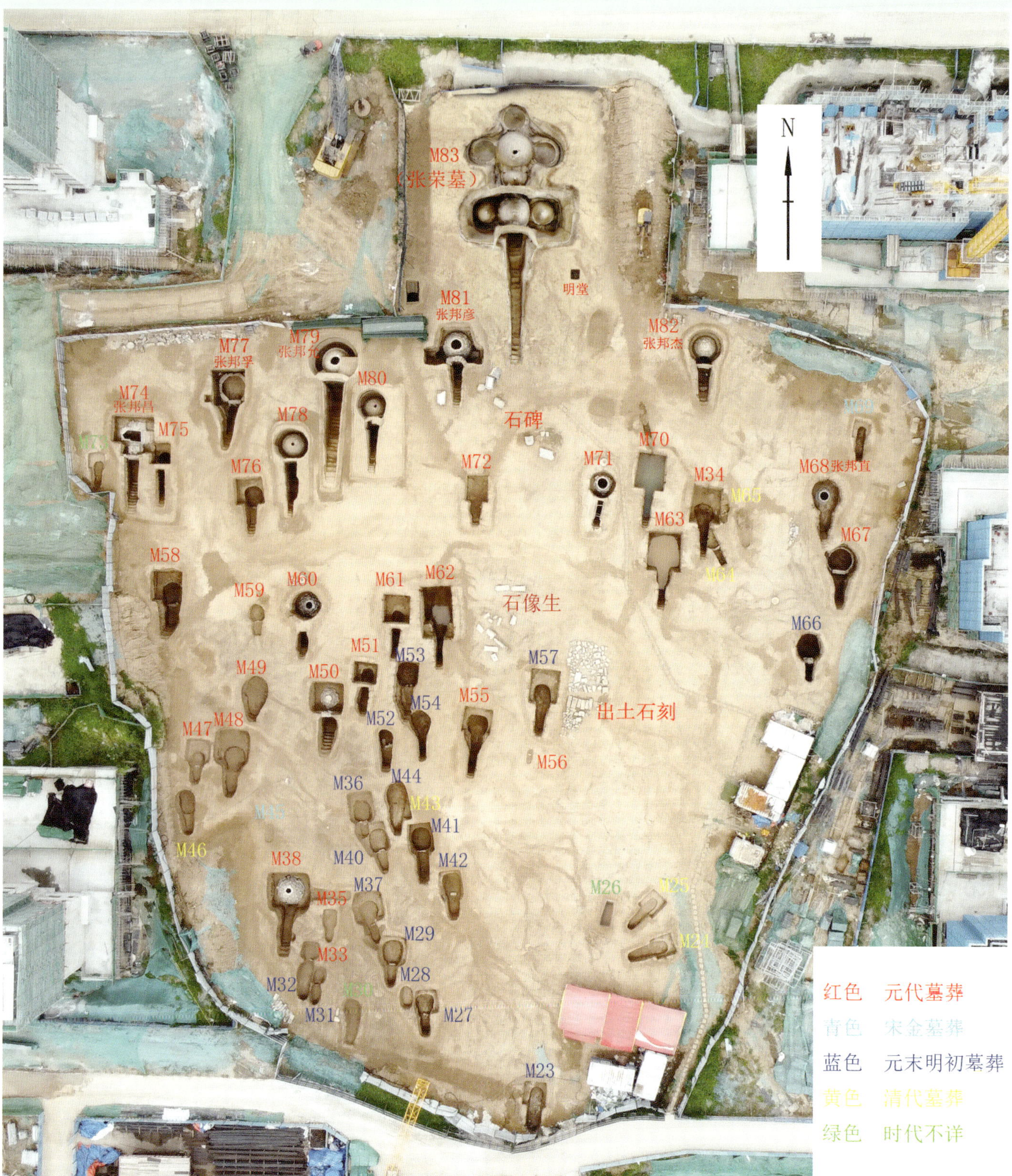

张荣家族墓地航拍图
Aerial Photo of Zhang Rong's Family Tombs

张荣墓前室
The Front Chamber of Zhang Rong's Tomb

张荣墓中室
The Middle Chamber of Zhang Rong's Tomb

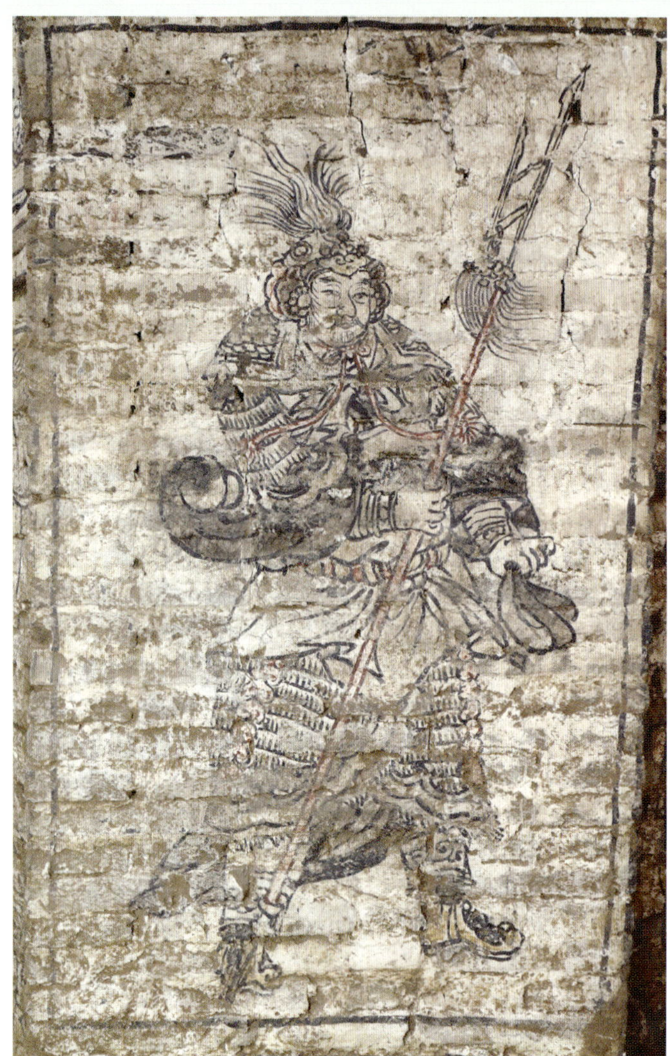

张荣墓后室门道东、西壁“持戟武士”图
Picture of the “Warrior Holding a Spear” on the East and West Walls of the Entrancc to Zhang Rong’s Tomb

白地褐彩双鱼纹瓷盆

Porcelain Basin with Double Fish Pattern in Underglaze Brown on White Ground

元（1206—1368）
济南王张荣家族墓地出土
济南市考古研究院藏
口径 46.78 厘米，高 13.7 厘米，底径 22.9 厘米

磁州窑瓷器。敞口，平折沿，斜腹内收，平底。器内口沿处绘 4 组黑色双线莲瓣纹，其下方饰两道弦纹，弦纹之间卷云纹，底部绘一饱满荷叶纹。

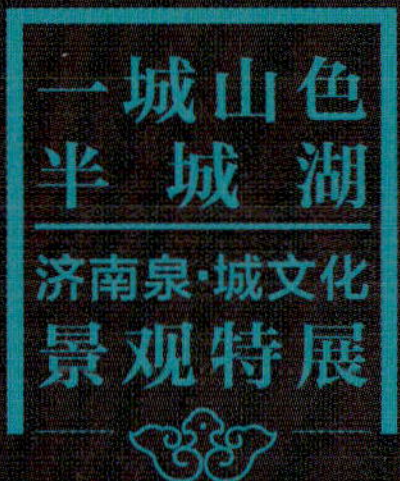

青白釉玉壶春瓶

Yuhuchun Vase with Bluish-white Glaze

元（1206—1368）
济南王张荣家族墓地出土
济南市考古研究院藏
口径 7.53 厘米，高 23.3 厘米

景德镇窑产品。施青白釉，器底及圈足内无釉，釉色均匀，细腻莹润，器身刻花卉纹，刀法流畅娴熟。

豆青釉玉壶春瓶

Yuhuchun Vase with Yellowish Pea Green Glaze

元（1206—1368）
济南王张荣家族墓地出土
济南市考古研究院藏
口径 6.51 厘米，高 26.9 厘米，底径 7.47 厘米

龙泉窑系产品。喇叭状侈口，细长颈，垂腹，圈足。施豆青釉，器底及圈足内无釉，釉色青翠，光洁纯净，温润如玉。

青白釉瓷匜（yí）

Bluish-white-glazed Porcelain Yi

元（1206—1368）
济南王张荣家族墓地出土
济南市考古研究院藏
长 17.2 厘米，口径 13.44 厘米，高 4 厘米

敞口，浅弧腹，平底。口一侧出槽状流，流下置一卷云形系，造型流畅。内外施青白釉，釉色均匀厚重。口沿和外底未经釉饰。

青釉瓷匜（yí）

Celadon-glazed Porcelain Yi

元（1206—1368）
济南王张荣家族墓地出土
济南市考古研究院藏
长 21.3 厘米，口径 16.2 厘米，高 5.52 厘米

敞口，圆唇，浅弧腹，平底。口一侧出槽形流。周身施青釉，釉色青翠，仅足底垫烧处无釉。上腹部饰弦纹两周，底部有一墨书。

青白釉瓷盘

Yuhuchun Vase with Blue-white glaze

元（1206—1368）
济南王张荣家族墓地出土
济南市考古研究院藏
口径 15.7 厘米，高 4.7 厘米，底径 5.1 厘米

敞口，圆唇，浅腹，圈足，内外施青釉。开片纹。盘心刻花卉纹。

青釉瓷盘

Celadon-glazed Porcelain Plate

元（1206—1368）
济南王张荣家族墓地出土
济南市考古研究院藏
口径 16.7 厘米，高 3.17 厘米，底径 8.19 厘米

敞口，浅腹，圈足，内外施青釉，足底无釉，釉色青翠莹润，釉面有凸起和气泡，盘内中心釉面开裂。盘内饰弦纹一周。

青花瓷盘

Blue and White Porcelain Plate

元（1206—1368）
济南王张荣家族墓地出土
济南市考古研究院藏
口径 15.4 厘米，高 1.78 厘米，底径 12.7 厘米

口沿绘卷草纹，盘内绘双鸳顾盼穿行于莲藻之间，暗含“连”“早”“缘”等吉祥寓意。

青花瓷碗

Blue and White Porcelain Bowl

元（1206—1368）
济南王张荣家族墓地出土
济南市考古研究院藏
口径 7.97 厘米，高 3.71 厘米，底径 3.1 厘米

内外均饰折枝花卉。青花色彩富有变化，浓淡相宜。胎体薄而致密。

青白釉花口瓷碗

Blue and white-glazed Porcelain Bowl with Flower Rim

元（1206—1368）
济南王张荣家族墓地出土
济南市考古研究院藏
口径 7.68 厘米，高 3.92 厘米，底径 3.31 厘米

花口，深腹，圈足，内外器壁有花瓣状凸起。胎壁较薄。器身施白釉，白中泛青。

青釉瓷碗

Celadon - glazed Porcelain Bowl

元（1206—1368）
济南王张荣家族墓地出土
济南市考古研究院藏
口径 7.28 厘米，高 3.55 厘米，底径 2.32 厘米

敞口，弧腹，圈足。内外施青釉。

白地褐彩龙凤纹瓷罐

White-ground Porcelain Jar with Brown-Decorated Dragon and Phoenix Patterns

元（1206—1368）
济南王张荣家族墓地出土
济南市考古研究院藏
口径 16.1 厘米，腹径 33.5 厘米，底径 12.3 厘米，高 32.6 厘米

上腹部为一龙两凤主题纹饰，龙昂首前伸，身姿矫健，足踩祥云，两凤一前一后，曲颈展翅，凤尾随风飘舞，为元代磁州窑类型的优秀作品。

奁（lián）式青釉瓷香炉

Double-series Jun-glazed Porcelain Jar

元（1206—1368）
济南王张荣家族墓地出土
济南市考古研究院藏
口径 13.66 厘米，高 7.1 厘米，底径 4.42 厘米

圆唇，直筒，圈足着地，外三兽蹄足悬空。腹部有凸起八卦纹，下腹部饰两圈弦纹。施青釉，釉色薄而均匀，清亮雅致。圈足及足底素烧无釉。

钧釉双系瓷罐

Double-series Jun-glazed Porcelain Jar

元（1206—1368）
济南王张荣家族墓地出土
济南市考古研究院藏
口径 14.62 厘米，高 14 厘米，底径 8.13 厘米

直口，圆肩，弧腹，圈足。口沿与肩上部有桥形双系。器身施蓝釉，口沿处釉层较薄，施釉不及底，釉内多细小气泡。釉面开片纹。

彩绘乐伎陶俑
Painted Pottery Figurine of Musician

元（1206—1368）
济南王张荣家族墓地张荣墓出土
济南市考古研究院藏
高 44.7 厘米，底长 11.9 厘米，底宽 10.6 厘米

人物包头髻，面容丰满，上身穿红色窄袖内搭，外搭对襟短袖坎肩，屈肘持笛于身前，下身裙裾及地，仅露足尖。

彩绘持壶陶俑
Painted Pottery Figurine carrying Pot

元（1206—1368）
济南王张荣家族墓地张荣墓出土
济南市考古研究院藏
残高 44.3 厘米，肩宽 9.8 厘米

人物身材高大，微屈膝站立，头戴冠，上身着内搭，外披短袖对襟坎肩，双手执壶于胸前，长衫及地。

彩绘陶俑

Painted Pottery Figurine

元（1206—1368）
济南王张荣家族墓地张荣墓出土
济南市考古研究院藏
高 48 厘米，底长 14.3 厘米，底宽 12.8 厘米，肩宽 13 厘米

人物头戴幞头，目光坚定，神色严肃，面周蓄有短须，身着圆领广袖长袍，腰系带，长袍及地，双手执物于胸前。

元彩绘侍女俑

Pottery Male Standing Figurine

元（1206—1368）
济南王张荣家族墓地张邦昌墓出土
济南市考古研究院藏
高 30.3 厘米，裙摆宽 11.6 厘米，底 11.3×10.9 厘米，肩宽 6.7 厘米

侍女头梳双环短髻于耳后，脸庞圆润，面带微笑，上身内着窄袖衫，外罩对襟短褂，双手屈肘于胸前，捧一方帕，帕上托一盘，盘上置一盏。

陶男立俑

Pottery Male Standing Figurine

元（1206—1368）
济南王张荣家族墓地张邦昌墓出土
济南市考古研究院藏
高 37 厘米，肩宽 7.8 厘米

人物形象为一中年男性，头戴幞头，面庞丰润，眼尾上挑，唇微张，面中有八字胡须，下颌蓄山羊胡，身着圆领窄袖长袍，下摆及膝，双手屈肘于胸前做握绳状，可能为牵马俑。

陶男立俑

Pottery Male Standing Figurine

元（1206—1368）
济南王张荣家族墓地张邦昌墓出土
济南市考古研究院藏
高 36.4 厘米，肩宽 8.4 厘米

人物面中与下颌蓄有短须，身穿斜襟长衫，腰系带，右手握于腰间，左臂垂于身侧。

陶男立俑

Pottery Male Standing Figurine

元（1206—1368）
济南王张荣家族墓地张邦昌墓出土
济南市考古研究院藏
高 34.2 厘米，肩宽 8.7 厘米

人物浓眉上挑，吊稍眼，双唇紧闭，神情严肃。右手握于腹前，左臂垂于腰侧。

山东首府（明清时期）

明洪武元年（1368），济南成为山东省会，是全省的政治、军事、经济、文化中心。明清两代，济南成为南北两京、华北与江淮两大平原之间的经济重镇，是全国重要的中心城市之一。作为山东首府，济南城内分布着衙署、官办学府、祭祀庙祠、宅院等各类建筑，特别是遍布城中的明代王府，对济南的城市形态、地域结构和土地规划与利用产生了深远影响。

The Capital of Shandong (Ming and Qing Dynasties)

In the first year of Hongwu of Ming Dynasty (1368), Jinan became the capital of Shandong Province and the political, military, economic, and cultural center of the province. During the Ming and Qing dynasties, Jinan emerged as a pivotal economic hub bridging the two capitals (Nanjing and Beijing), the North China Plain, and the Jianghuai Plain, solidifying its status as one of China's foremost central cities. As the capital of Shandong Province, Jinan's urban landscape was shaped by a diverse array of architectural complexes, including administrative offices, state-run academies, sacrificial temples and residential compounds. Notably, the widespread presence of Ming princely residences profoundly influenced the city's morphology, spatial organization, and land-use planning.

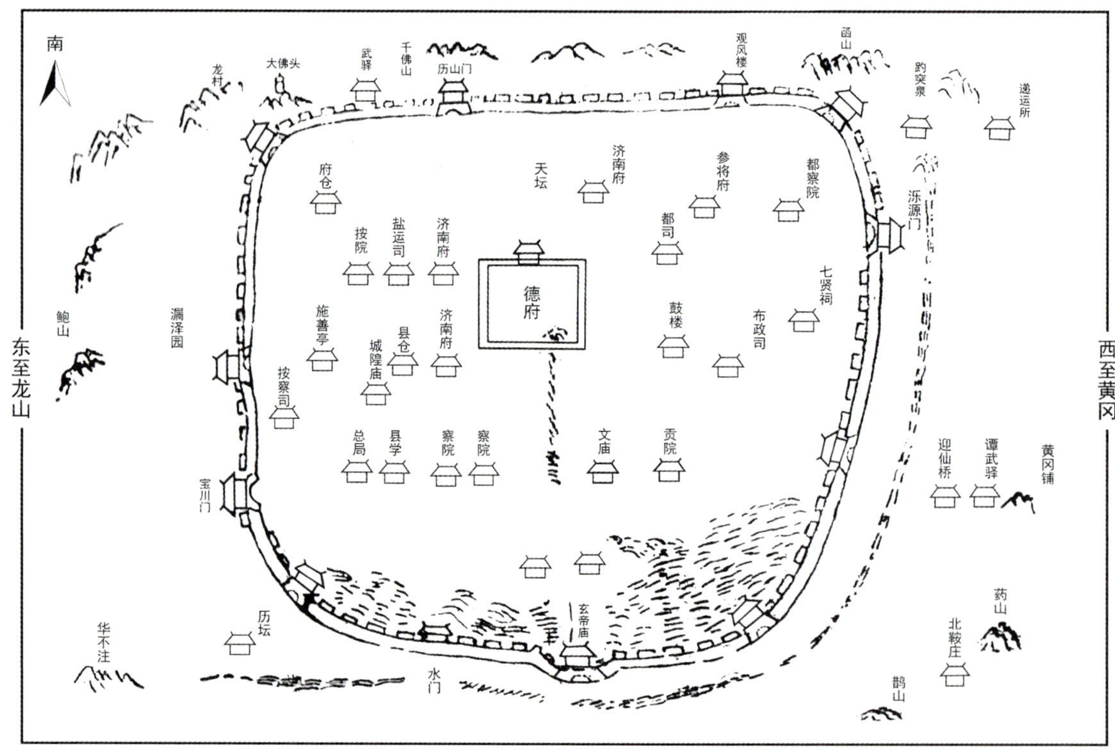

明末济南城图
The Map of Jinan City in the late Ming Dynasty

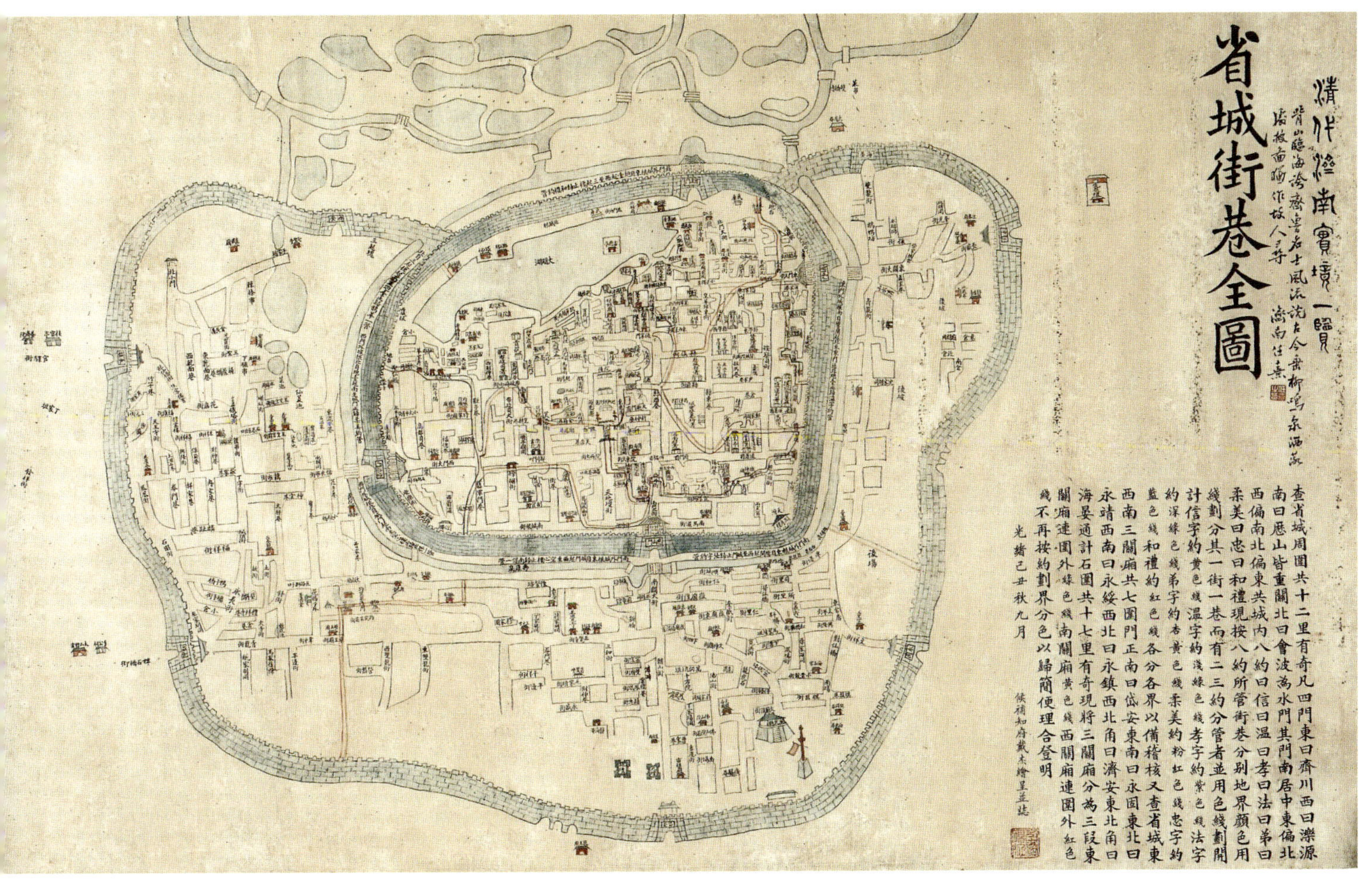

《省城街巷全图》

A Complete Map of the Streets and Alleys in the Provincial Capital

1902 年，新任山东巡抚周馥在书肆中找到 1898 年间旧版地图，将其重新分区染色编成新图，名为《省城街巷全图》，现藏于山东省图书馆，与 1841 年《济南府城并各衙门图》相比，此图明显变化是在城外修建有外郭，城区向东、西、南三面拓展超过一倍。

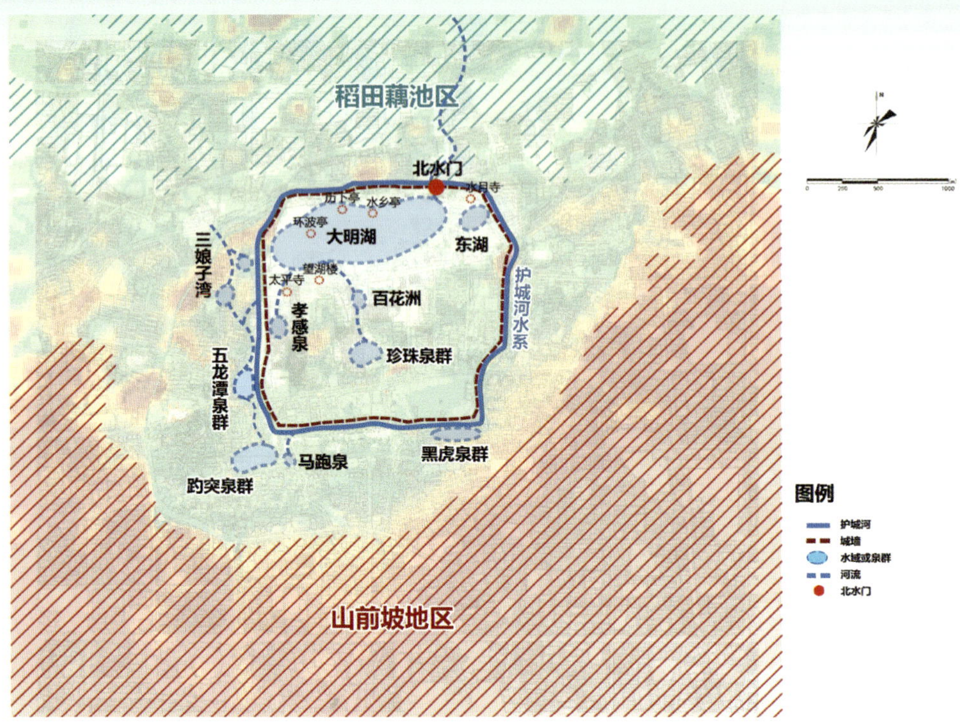

明代水环境示意图
Schematic Diagram of the Water Environment in the Ming Dynasty

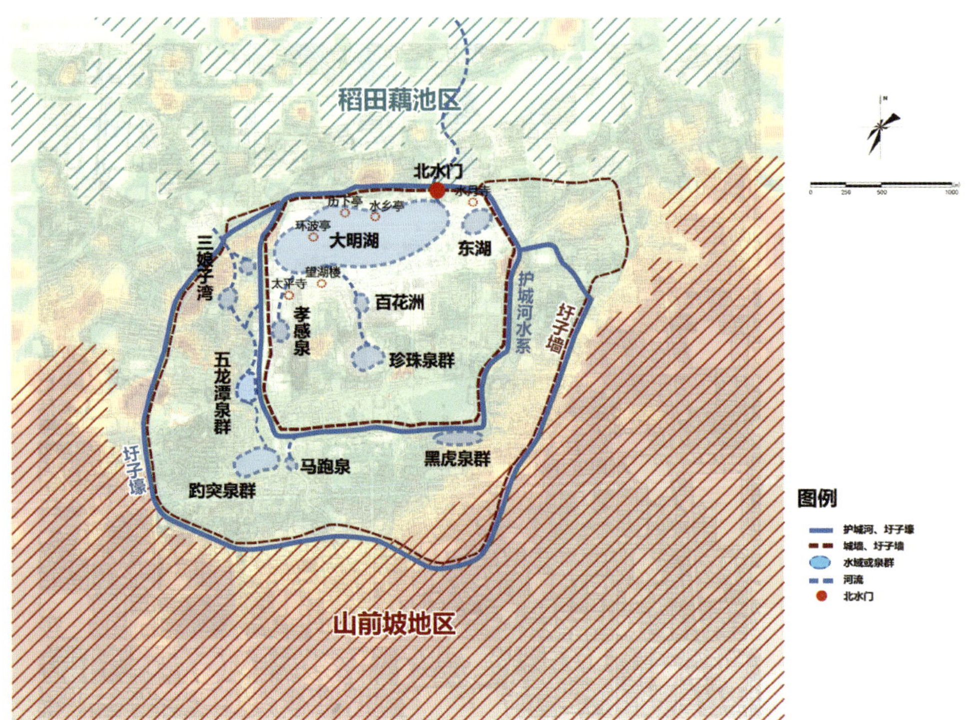

清代水环境示意图
Schematic Diagram of the Water Environment in the Qing Dynasty

明代济南王府、衙署、寺庙分布图
Distribution Map of Jinan Princely Residences, Administrative Offices and Temples in the Ming Dynasty

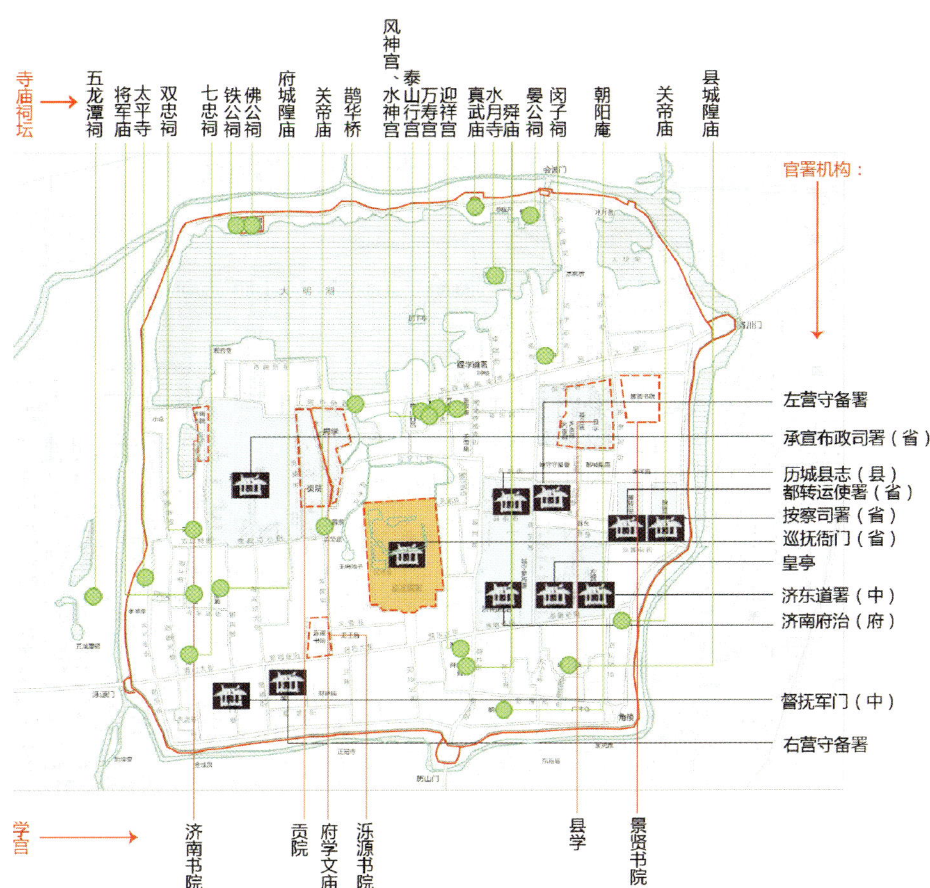

清代济南衙署、学宫、寺庙分布图
Distribution Map of Jinan Government Offices, Academic Palaces, and Temples in the Qing Dynasty

藩屏齐鲁

明朝建立后，为了巩固统治，明太祖朱元璋实行了封建诸王之制，广封诸子为王，镇守要塞、备侮御边，形成拱卫帝室的藩屏。在山东地区，先后分封了齐、鲁、汉、德、衡、泾六王，其中德王就封于济南。

明德王府

德王府位于珍珠泉畔，亦称“德藩故宫”，建于前朝济南王张荣的府邸旧址上，规模宏大，建筑豪华，为明代济南城中最大的建筑群。它东至县西巷，西至芙蓉街，南至泉城路，北至后宰门街。据《历城县志》载：“德府，济南府治西，居会城中，占三之一。”清兵占领济南后，德王府的建筑大都被焚毁，后在此建山东巡抚院署。

Qilu of Vassal State

After the establishment of the Ming Dynasty, in order to consolidate his rule, Zhu Yuanzhang, founder of the Ming dynasty, implemented the feudal system of various kings, widely enfeoffed his sons as kings, guarded fortresses, prepared to bully the border, and formed a fiefdom screen to guard the imperial court. In the Shandong region, the six kings of Qi, Lu, Han, De, Heng, and Jing were successively enfeoffed, among which the king of De was enfeoffed in Jinan.

Residence of Prince De of the Ming Dynasty

The Residence of Prince De is located near the Pearl Spring, also known as the “De Princely Forbidden City”. It was lavishly built on the former site of the former residence of Zhang Rong, the Duke of Jinan in the previous dynasty. With a grand scale and luxurious architecture, it was the largest architectural complex in Jinan during the Ming Dynasty. It extends east to Xianxi Lane, west to Furong Street, south to Quancheng Road, and north to Houzaimen Street. According to the *Li Cheng County Annals*, “Residence of Prince De of the Ming Dynasty was located west of the Jinan Prefectural seat, occupying one-third of the city area.” After the Qing army occupied Jinan, most of the buildings in the Residence of Prince De were burned down, and the Shandong Imperial Commissioner’s Yamen was later built here.

明代德王府范围示意图
Schematic Diagram of the Scope of the Residence of Prince De in the Ming Dynasty

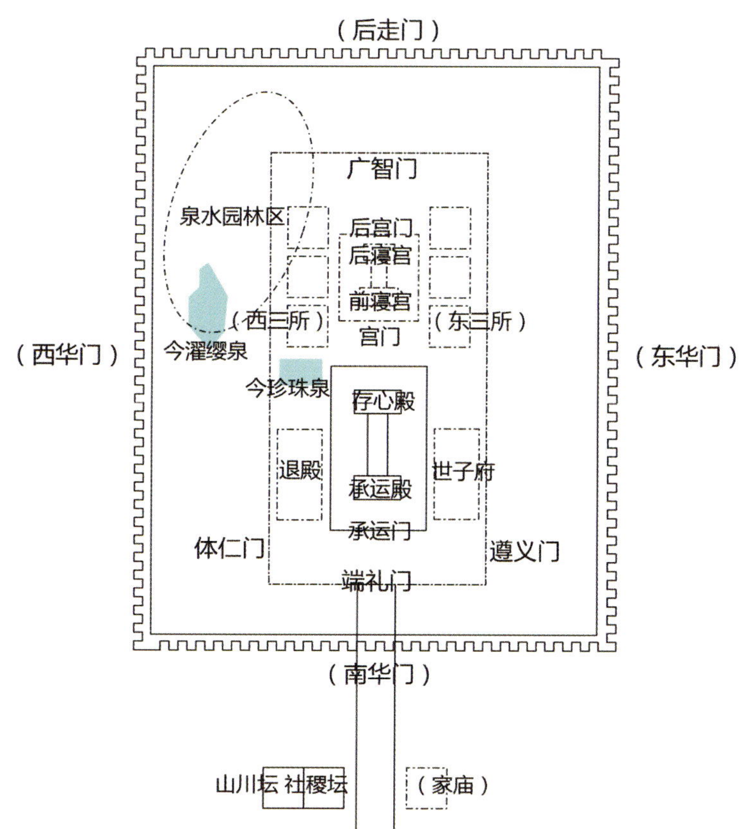

明代德王府宫城空间布局示意图
Schematic Diagram of the Spatial Layout of the palace city of the Residence of Prince De in the Ming Dynasty

明德王墓群

明德王墓群是目前已知规模宏大、形制清楚、保存最为完整的明代亲王家族墓地，位于长清区五峰山街道青崖寨山以南、以东，自东向西排列有7座德王墓。7座墓中除M5，其余均建有规模宏大的地面设施，圈筑内外陵园，构成内外城郭。陵园内置以享殿和配殿，神道两侧原有碑碣、石像生等。

Tombs of Princes De of the Ming Dynasty

Tombs of Princes De of the Ming Dynasty are currently known as the largest, most clearly shaped, and well preserved tombs of Prince Mingde in the Ming dynasty. It is located south and east of Qingyazhai Mountain in Wufengshan Street, Changqing District, with seven Prince Mingde tombs arranged from east to west. Among the 7 tombs, except for M5, all others have large-scale ground facilities, enclosing the inner and outer cemeteries, forming the inner and outer city walls. The mausoleum complex contained a Sacrificial Hall and Ancillary Halls. Flanking the Spirit Way were originally stone steles, stone statues of figures and animals, and other ceremonial sculptures.

⭑全国重点文物保护单位
National Key Cultural Relics Protection Unit

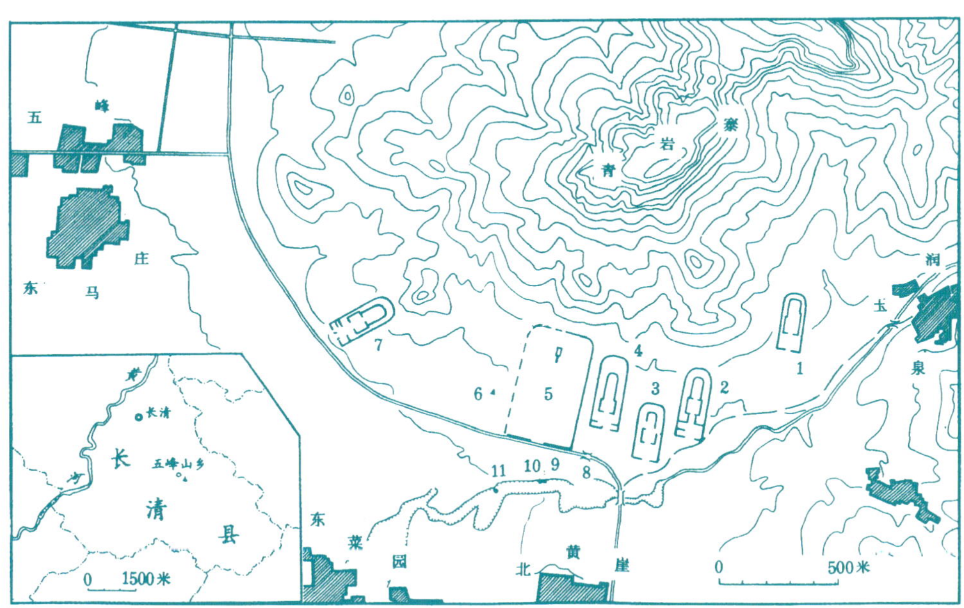

明德王墓群位置及墓葬分布示意图
Schematic Diagram of the Location and Distribution of the Tombs of Princes De in the Ming Dynasty

德庄王墓

德庄王墓（M4）墓室位于享殿正后方，是7座墓中规模最大的一座。墓内出土德庄王朱见潾和妃刘氏墓志各一。通过性别与年龄鉴定，结合棺椁等级，推知东殿葬德庄王朱见潾，西殿葬王妃刘氏，前殿葬济宁安僖王朱祐柃。

The Tomb of Prince Dezhuang

The Tomb of Prince Dezhuang (M4) is located directly behind the palace and is the largest of the seven tombs. The epitaphs of Zhu Jianlin, Prince of Dezhuang, and his Concubine surnamed Liu were unearthed in the tomb. Through gender and age identification, combined with the level of the coffin, it is inferred that Zhu Jianlin, Prince of Dezhuang was buried in the Eastern Hall, while his Concubine surnamed Liu was buried in the Western Hall. Zhu Youxun, Prince Anxi of Jining was buried in the front hall.

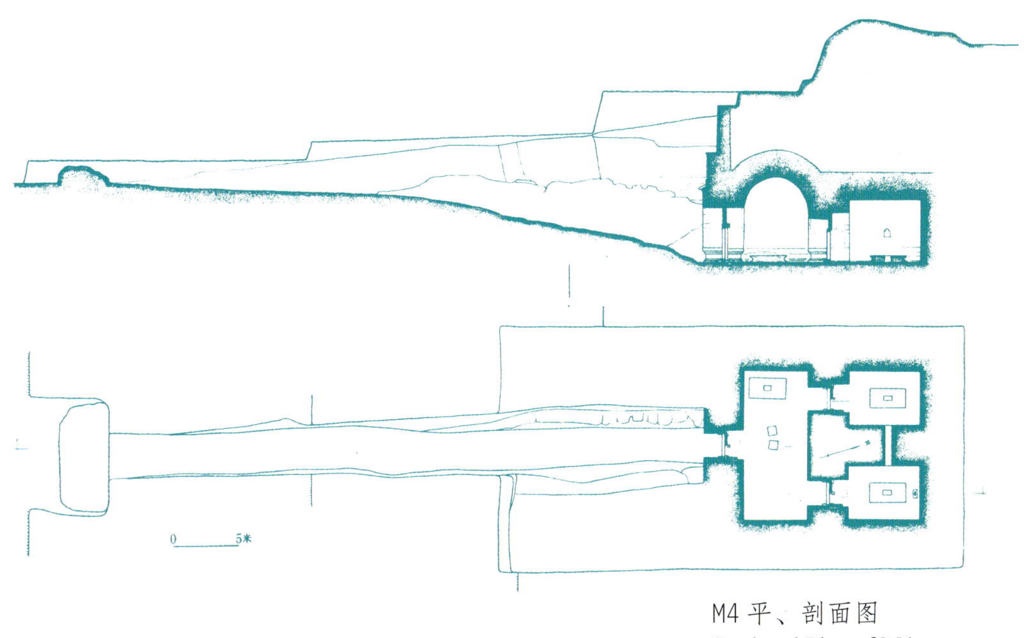

M4 平、剖面图
Sectional Plan of M4

德莊王
壙誌銘

德庄王墓志盖拓本
Rubbing of Epitaph Cover of the Tomb of Prince Dezhuan

德庄王墓志拓本
Rubbing of Epitaph of the Tomb of Prince Dezhuang

刘妃墓志拓本
Epitaph of Imperial Concubine Surnamed Liu

明代郡王府

明代亲王由嫡长子世袭，其余诸子则封为郡王，并在亲王所在地分建郡王府。德王袭七代八位，其子孙先后有十五人受封郡王，到明末尚有八位郡王，相应就有八座郡王府，遍布济南城内：西门内有泰安王府、临朐王府；南门内有宁海王府、宁阳王府；尹家巷有临清王府；县庠西有纪城王府；还有位于府馆街的嘉祥王府、清平王府。这些郡王府几乎占据了济南府城的半壁江山。

The Prince's Mansion of the Ming Dynasty

In the Ming Dynasty, the princely title was inherited by the eldest legitimate son through primogeniture, while other sons were enfeoffed as commandery princes. Separate commandery prince residences were established within the jurisdiction of their father's princedom. The Prince of De witnessed seven generations with eight successions. Fifteen of his descendants were enfeoffed as commandery princes, of whom eight retained their titles of commander princes by the late Ming Dynasty. Correspondingly, eight commandery prince residences were distributed across Jinan City: Prince of Tai'an and Prince of Linqu resided within the West Gate; Within the South Gate was the place of residence of Prince of Ninghai and Prince of Ningyang; Prince of Linqing resided in Yinjia Lane; In the west of the county yamen was the place of residence of the Prince Jicheng; Palaces of the Prince of Jiaxiang and Prince of Qingping were resided in the Fuguan Street. These commandery prince residences almost occupied half of the Jinan city.

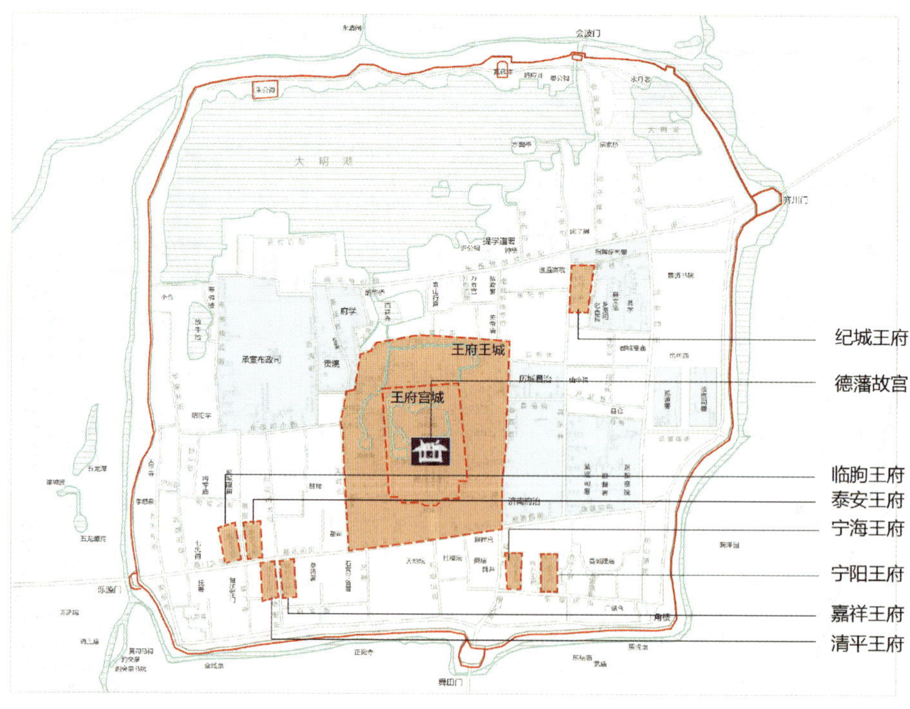

明代晚期济南城各王府位置图
Location Map of Various Commander Prince Residences in Jinan City during the Late Ming Dynasty

龙纹镜

Mirror with Dragon Patterns

元至元四年（1338）
历城明纪城王墓出土
济南市博物馆藏
直径 22.5 厘米，厚 0.8 厘米

此镜圆形，圆钮，钮外一方框，框内铸有“至元四年”阳文，方框左右各饰一火珠纹，钮座上下用浮雕技法各饰一飞龙。飞龙弓背曲腰，昂首盘旋，张口露齿，一爪前伸，作欲吞食火珠状。二龙周边间饰花卉纹、荷叶纹等。

宁阳王府和宁海王府

济南古城东南隅的宽厚所街明代郡王府遗址，发掘出明代宁阳王府和宁海王府，其中前者布局完整，后者仅残留院落东北角。宁阳王府院落整体坐北朝南，占地面积约12000平方米，显示出王府规模之宏大。为了防止发生地下水位过高、遇雨宣泄不及的情况，院内地面经整体垫高、整平，明显高于院墙外的同时期地面。

Commander Prince Residences of the Prince of Ningyang and the prince of Ninghai

The Historic Site of commander prince residence of the Ming Dynasty on Kuangshuo Street in the southeast corner of Jinan Ancient City have been excavated, and commander Prince Residence of Ningyang and commander Prince Residence of Ninghai of the Ming Dynasty were discovered. The former has a complete layout, while the latter only remains in the northeast corner of the courtyard. Commander Prince Residence of Ningyang faces north and south, covering an area of approximately 12,000m^2, demonstrating the grandeur scale of the commander prince residence. In order to prevent conditions where the groundwater level is high and there is insufficient rain release, the ground level inside the courtyard has been raised and leveled as a whole, significantly higher than the ground level outside the courtyard walls at the same time.

宁阳王府遗址航拍图（上为北）
Aerial View of the Historic Site of Residence of Commander Prince of Ningyang (North in the Upper Direction)

宁阳王府东墙中段残存最高处（自东北向西南）
The Highest Remaining Part of the Middle Section of the East Wall of the Residence of Commander Prince of Ningyang (from Northeast to Southwest)

宁阳王府北院墙
North Courtyard Wall of the Residence of Commander Prince of Ningyang

宁阳王府东路院中进院落之正房 F9（由东向西）
Principal Room F9 in the Central Courtyard in the Courtyard on Wangfu East Road, Ningyang (From East to West)

宁阳王府东路院最南进院落（由北向南）
The south yard of the courtyard on the east road in Ningyangwang Palace（from north to south）

宁阳王府东路院最南进院落之东厢房 F23（由北向南）
East Wing Room F23 in the Southernmost Courtyard of the Courtyard on Wangfu East Road, Ningyang (from north to south)

宁阳王府 F23 西侧甬道
West Side Corridor of F23 in the Residence of Commander Prince of Ningyang

宁阳王府后阁楼（王府内最大的单体建筑）
The Rear Attic of Residence of Commander Prince of Ningyang (the Largest Single Building within the Residence)

宁阳王府内残留大型柱础（由东向西）
Large Column Foundations Remain in Residence of Commander Prince of Ningyang (From East to West)

宁阳王府北院水井
Water Well in the North Courtyard of the Residence of Commander Prince of Ningyang

宁海王府内水井
Water Well within Residence of Commander Prince of Ningyang

排水设施

宁阳王府院内具备完善的排水设施，整个系统由房屋前后的散水、明渠式水道、覆盖石板的暗渠式下水道、院墙上的排水口组成。其中，暗渠共清理出 10 条，总长度达 400 多米。

Drainage Facilities

Residence of Commander Prince of Ningyang has complete drainage facilities in its courtyard. The entire system consists of water apron before and after the house, open channel-type waterways, culvert-type sewers covered with slate, and drainage outlets on the courtyard wall. Among them, a total of 10 underground culverts were cleared, with a total length of over 400m.

宁阳王府东墙南段内侧散水（由北向南）
Aerial View of the Historic Site of Residence of Commander Prince of Ningyang (North in the Above Direction)

宁阳王府东墙中段内侧暗渠下水道（由东北向西南）
Culvert Sewer on the Inner Side of the Middle Section of the East Wall of Residence of Commander Prince of Ningyang (From Northeast to Southwest)

宁阳王府暗渠 G8（由东北向西南）
Culvert G8 of Residence of Commander Prince of Ningyang (From Northeast to Southwest)

宁阳王府西北角出水口
Water Outlet of Northwest Corner of Residence of Commander Prince of Ningyang

中间 3 个石桩，既能保证水源流出，又具有防盗功能（由北向南）

宁阳王府北墙出水口和基址
Water Outlet and Foundation Site of the North wall of Residence of Commander Prince of Ningyang

宁阳王府东北角出水口（由北向南）

Water Outlet of the North wall of Residence of Commander Prince of Ningyang (From North to South)

石砌水池

在宁阳王府大门南侧偏西约 20 米处，发现一座长方形石砌水池，内径长 12 米、宽 7 米、深 2.3 米，由规整的长方形条石砌筑，底铺石板。池底石板下中部填有约 1 米厚不规则石块，而中部的 3 块石板上各有两个半月形孔，据此推测，该池可能为一砌在泉水之上并具有净化作用的水池，或为王府用水来源之一。

Stone-built Water Tank

A rectangular stone-built water tank was discovered approximately 20m west of the south gate of Residence of Commander Prince of Ningyang. It has an inner diameter of 12m, width of 7m, and depth of 2.3m, and is constructed with regular rectangular stone strips and paved with stone slabs at the bottom. The middle part of the stone slabs at the bottom of the tank is filled with irregular stones about 1m in thickness; and each of the three slabs in the middle has two half-moon holes. It is speculated that the tank may be a water tank built on top of the spring water and has a purifying effect, or it is one of the sources of water for the commander prince residence.

宁阳王府南侧水池遗迹
The Remains of the Water Tank on the South Side of the Residence of Commander Prince of Ningyang

青釉琮（cóng）式瓷瓶

Celadon-glazed Cong-shaped Porcelain Vase

宋（960—1276）
宽厚所街遗址出土
济南市考古研究院藏
高 25.8 厘米，口径 8 厘米，底径 8.2 厘米，瓶身边长 9.9 厘米

影青瓷。仿琮形，方柱体高筒形，外方内圆。外面由竖槽分为左右两个部分。每个部分又被凸棱分成8节。

黄釉花口瓷瓶

Golden-glazed Porcelain Vase with Foliate Rim

金（1115—1234）
宽厚所街遗址出土
济南市考古研究院藏
口径 9 厘米，高 23.4 厘米，底径 9.5 厘米

莲花形瓶口，细长颈，有五圈内凹弦纹。通体施黄釉，圈足边缘及内底无釉，露出赭黄色内胎。素面，无纹饰。整器造型优雅，端庄秀气，比例协调。

衙署祠庙

珍珠泉畔清山东巡抚衙门旧址

珍珠泉是济南“四大名泉”之一，因它位于古城中心且泉眼众多、涌姿独特，成为历代文人墨客观赏、吟咏的佳景，也成为济南“七十二名泉”中最早被圈入园林住宅的名泉之一。珍珠泉畔及其周围区域，先后被建成知济南府第（金、元）、德王府（明）、山东巡抚衙门（清）。随着王朝更迭和战火纷争，现只有清巡抚院署大堂仍矗立在珍珠泉大院内。

Administrative Office and Ancestral Temples

Former Site of the Qing Shandong Provincial Governor's Yamen by the Pearl Spring

Pearl Spring, one of Jinan's "Four Eminent Springs", has been a celebrated cultural landmark since antiquity. Nestled in the historic city center, its abundant spring holes bubble up in distinctive rhythms, inspiring literati and poets through dynasties to gather here for contemplation and verse. Notably, it ranks among the earliest of Jinan's "72 Celebrated Springs" to be enclosed within a private garden residence, reflecting its enduring aesthetic and symbolic value. The Pearl Spring and its surrounding areas were successively enclosed into the Residence of Jinan Prefecture (in the Jin and Yuan Dynasties), Residence of Prince De (in the Ming Dynasty), and the Shandong Governor's Yamen (in the Qing Dynasty). With the change of dynasties and wars, now only principal hall of the Imperial Commissioner's Yamen still stands in the courtyard of Pearl Spring.

1928 年山东省督军署（清山东巡抚衙门旧址）航拍图
Aerial Photo of Shandong Provincial Military Governor's Office in 1928 (The Former Site of Shandong Imperial Commissioner's Yamen of the Qing Dynasty)

珍珠泉历史影像
The Historical Image of Pearl Spring

清巡抚院署大堂

清巡抚院署大堂，名承运殿，由山东巡抚周有德于康熙五年（1666）在原明德王府中心部位重新修建而成。大堂面阔五间，进深四间，建筑面积620余平方米，是济南现存唯一的衙门建筑。

Principal Hall of the Imperial Commissioner's Yamen of the Qing Dynasty

The principal hall of the Qing Governor's Office, named Chengyun Palace, was rebuilt by Shandong Imperial Commissioner Zhou Youde in the fifth year of Kangxi (1666) in the center of the original central location of the residence of Price De in the Ming Dynasty. The principal hall spans five bays in width and four bays in depth, covering a floor area of over 620m^2. It is the only existing yamen building in Jinan.

1928年山东省督办公署汽车库（清巡抚院署大堂旧址）
Shandong Provincial Governor's Office Garage in 1928 (the Former Site of Principal Hall of the Imperial Commissioner's Yamen of the Qing Dynasty)

清巡抚院署大堂北立面
the North Elevation of Principal Hall of the Imperial Commissioner's Yamen of the Qing Dynasty

百花洲泰山行宫遗址

泰山行宫是供奉泰山女神碧霞元君的宫庙。百花洲泰山行宫是历城区开创年代最早的泰山行宫之一，始建于明正德十一年（1516），明清及民国多次重修，20 世纪 90 年代废弃，延续 400 余年。现存泰山行宫遗址可分为早、中、晚三期，推测分别为明代创修期、清代重修期、民国重修期。这两次重修均是为了防止室内潮湿、内涝而抬升了室内地面，表明地下水位渐趋上升，泉水生态环境良好。

The Historic Site of Baihuazhou Mount Tai Imperial Palace

Mount Tai Imperial Palace is a temple dedicated to Bixia Yuanjun, the Goddess of Mount Tai. Baihuazhou Mount Taishan Palace is one of the earliest Mount Tai Palace in Licheng District. It was built in the 11th year of Zhengde (1516) of the Ming Dynasty. It was rebuilt for many times in the Ming Qing Dynasties and the period of the Republic of China. It was abandoned in the 1990s and lasted for more than 400 years. The existing Historic Site of Mount Tai Palace can be divided into early, middle and late periods, which are speculated to be the construction and the restoration period of the Ming Dynasty, and the renovation period of the Qing Dynasty, and the restoration period of the Republic of China. The renovations in the Qing Dynasty and the period of the Republic of China were both carried out to prevent indoor moisture and waterlogging, which raised the indoor floor, indicating that the groundwater level is gradually rising and the ecological environment of the spring water is good.

泰山行宫遗址航拍照
Aerial Photo of the Historic Site of Baihuazhou Mount Tai Imperial Palace

0 20厘米

"新建泰山行宫碑记"碑额拓片
the Rubbing of Stele Heading of "Newly Built Mount Tai Imperial Palace Inscription"

0 20厘米

"重修泰山行宫碑记"碑额拓片
Inscription Rubbings of "Newly Built Mount Tai Imperial Palace Inscription"

陶脊兽

Pottery Ridge Beast (Pottery Decorative Components on the Roof Ridge of the Building)

清（1644—1911）
百花洲泰山行宫遗址出土
济南市考古研究院藏

脊兽为建筑构件，施于合脊筒瓦之上。

府学文庙

文庙不仅是古代重要的祭孔场所，也是组织科举考试的重要机构。济南府学文庙与曲阜孔庙、南京六合文庙、苏州文庙并称“中国四大文庙”。府学文庙是济南市区现存历史最早、规模最大的古建筑群之一，始建于宋熙宁年间（1068—1077），元末倾塌，明洪武二年（1369）重建，清代多次修葺，基本保持了明代文庙的规模和布局。

Fuxue Confucian Temple

Prefectural Confucian Temple was not only an important place for worshiping Confucius in ancient times, but also an important institution for organizing the imperial examination. Jinan Fuxue Confucian Temple, Qufu Confucian Temple, Nanjing Liuhe Confucian Temple, and Suzhou Confucian Temple are collectively known as the “four major Confucian temples in China”. The Fuxue Confucian Temple is one of the earliest and largest existing ancient architectural complexes in the urban area of Jinan. It was built during the Xining period of the Song Dynasty (1068-1077), collapsed at the end of the Yuan Dynasty, and was rebuilt in the second year of the Hongwu period of the Ming Dynasty (1369). It was repaired for many times in the Qing Dynasty, basically maintaining the scale and layout of the Confucian Temple of the Ming Dynasty.

* 山东省文物保护单位
Cultural Relics Protection Unit of Shandong Province

府学文庙
Fuxue Confucian Temple

济南府学文庙是典型的围绕泉池组织空间的庙宇建筑，反映了泉水作为公共资源的属性。

仿宋哥釉铁花钵式炉

Iron Oxide Floral Bowl-shaped Incense Burner Imitating the Song Dynasty's Geware Glaze

明（1368—1644）
历城明代赵玹墓出土
济南市博物馆藏
高 11 厘米，口径 18.8 厘米，底径 12.5 厘米，腹围 61 厘米

通体施豆青釉，釉面呈深浅不一的网状开片。颈部露胎，饰卷草花纹一周，并堆贴两狮面耳。口部、颈部和足部呈酱色，为仿制宋代哥窑的紫口、铁花、铁足型器。

青花花蝶纹梅瓶

Blue-and-White Porcelain Meiping with Floral and Butterfly Patterns

明（1368—1644）
历城明代赵玹墓出土
济南市博物馆藏
左：口径 6.2 厘米，底径 9.9 厘米，高 27.3 厘米，腹围 41 厘米
右：口径 6.3 厘米，底径 9.75 厘米，高 27 厘米，腹围 42 厘米

瓶为一对，形制相同。胎质白色，通体白釉，口沿施赭黄釉。器体为白釉青花绘蝴蝶、花叶纹。白釉厚润，青花纹饰简单清雅。

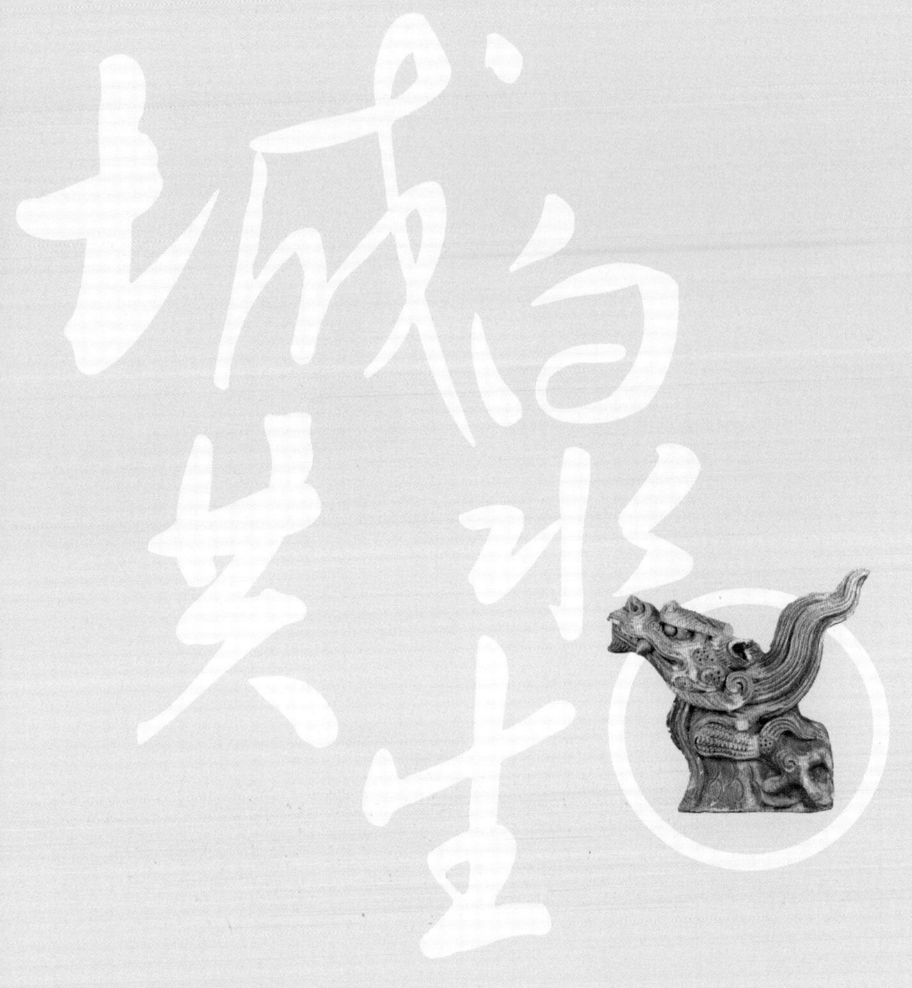

第三章 城泉共生

人文与自然交融的文化景观

泉水孕育了“泉城”，深刻影响着济南的城市营造智慧、名士文化追求和地域生活传统，推动了济南的城市发展。古人独具匠心，引水成渠，汇溪成湖，将山、泉、湖、河与城巧妙结合，织就了一幅秀美天成的“泉城”画卷；泉水因其清澈、甘冽的特性，被视为自然之美的象征，成为历代文人咏赞的对象；泉水也成为济南人生活中不可或缺的一部分，城依泉神韵倍增，泉依城源远流长。

Chapter III. City & Springs Symbiosis: Cultural Landscape of the Harmonious Integration between Humanities and Nature

Spring water has nurtured the “Spring City” and profoundly influenced the urban construction wisdom, pursuit of literati culture and regional living traditions of Jinan, and promoted the urban development of Jinan. The ancients had a unique and ingenious approach, diverting water into canals and merging streams into lakes. They ingeniously combined mountains, springs, lakes, rivers and cities to weave a beautiful and natural “Spring City” painting. Spring water, due to its clear and sweet characteristics, is regarded as a symbol of natural beauty and has become an object of praise for literati throughout history. Spring water has also become an indispensable part of the lives of people in Jinan. The charm of the city is multiplied by the spring water, and the spring water has a long history.

8—11 世纪：咏赞甘泉

北宋文学家曾巩在《齐州二堂记》中，叙述了『齐多甘泉』的缘由和当时趵突泉喷涌之情状，文中引用当地士人的称谓，第一次将『泺』正名为『趵突泉』，从此，这一古老的泉池便以『趵突泉』之名传诵千古。

在曾巩的另一名篇《齐州北水门记》中，首次记述了济南名泉的数目，大致与稍后的李格非《历下水记》所述的『三十余所』相合，这说明在北宋时期，济南尚无『七十二泉』之说。

12—14 世纪：名泉评鉴

金代，济南民间树立《名泉碑》，列举了济南七十二名泉。自此，便产生了济南『七十二泉』之说。中国古人习惯把『七十二』作为一个『成数』来使用，以表示某一事物数量之多，因此，世人便以『七十二名泉』之说来称赞古城济南泉水之多。据调查统计，济南市辖区范围内的名泉总数多达 950 处，这在国内乃至世界上，都是绝无仅有的。

元代地理学家于钦在《齐乘》中，把《名泉碑》内容原封不动地全部记录在书中，并逐一注出各泉在元时的所在地。

16–19 世纪：泉水文汇

明崇祯年间（1628—1644），刘敕在《历乘·舆地考》中系统地记述了明中后期济南名泉的概况，并将当时流传的著名咏泉诗文完整地收入书中，如晏壁的《济南七十二泉诗》、张珣的《七十二泉总咏》等，使这些珍贵的文学作品流传至今。清代郝植恭的《济南七十二泉记》记录了济南『七十二泉』的另一个版本。

在泉水环境影响下形成的『济南泉·城文化景观』起源于公元3世纪，发展于公元8—11世纪，成型于公元12世纪，繁荣于公元16—19世纪，长期演进持续至今。

○ 商代：甲骨文中的『泺』

1915年罗振玉在《殷墟书契考释》一书中发现甲骨文的『泺』字，将济南有文字记载的历史自春秋时期向前推进800年。

○ 公元前7世纪：泺水会盟

史书上最早关于泉水的记载，见于《春秋·桓公十八年》，其中有『公会齐侯于泺』的记录，记述了公元前694年鲁桓公与齐襄公在泺水会盟之事。泺水之源，就是现在的趵突泉。

○ 3世纪：林泉隐逸

魏晋以来隐逸文化发展，清谈之风盛行，文人雅士疏别功名利禄和荣华富贵，追求内心的安宁，关注世间万物的玄机与自然风物的百态，形成『山水林泉』文化。济南古城由于泉水环境特征突出，『泉』成为审美的重要要素。

○ 6世纪：水经著录

北魏时期，地理学家郦道元在《水经注·济水卷》中详细记述了当时的趵突泉、大明湖、净池（今五龙潭一带）、历水（今舜井）、流杯池（今珍珠泉、曲水亭一带）、华泉、百脉泉等几处济南名泉的状况，其中写道：『泺水出历（城）县故城西南，泉源上奋，水涌若轮。』这句话生动传神地描述出享有『群泉之冠』誉称的趵突泉的景象，并留下了『寰中之绝胜，古今之壮观』的赞咏。

纳泉为城

群泉环绕的济南古城一带，是人类先民难得的理想居住地。自古以来，丰沛的泉水就是济南居民生产、生活用水的重要来源。考古发现表明，济南古城“家家泉水、户户垂杨”的景观风貌有着悠久的历史。基于冷泉集中出露的特殊自然环境，济南古城形成了“纳泉而为”的选址、规模、布局及城垣形制，建立起“导—蓄”结合的城市水利体系，并发展出适应多水环境的建筑防水技术，成为中国北方罕见的山水园林城市。

枕水而居

利用天然水源是古人赖以生存的自然本领。济南古城内外的泉眼星罗棋布、水源充沛，吸引了古人在此定居生活，促进了生产和经济社会发展。古城内考古发现了数量众多、分布密集的古井，以及引水的沟渠和顺应泉渠而建的古代道路，这些发现是济南先民适应泉水环境、充分利用泉水资源的实证。枕水而居不仅是一种自然地理上的选择，也塑造了济南的文化传统和生活智慧。

高都司巷遗址的井渠和道路

高都司巷遗址位于古城西门附近，其年代上至春秋战国，下至明清时期，考古发掘揭露出路面、水井、窑址、窖藏等遗迹。遗址内共发现水井 40 余口，各水井之间相距不远，大部分距离两三米。在有限的地域内分布着如此多的水井，是十分罕见的。可以想见，早在 2000 多年前，这一地区便是一个水源旺盛、人口密集的地方。

Enfolding Springs as the City's Soul

The area of the ancient city of Jinan surrounded by springs is a rare and ideal place for human ancients to live. Since ancient times, abundant spring water has been an important source of water for the production and domestic use of Jinan residents. Archaeological discoveries indicate that the landscape of "spring water and poplar trees hanging in every household" in the ancient city of Jinan has a long history. Based on the unique natural environment where cold springs are concentrated and exposed, the ancient city of Jinan has formed a site selection, scale, layout and city wall structure for receiving springs. It has established an urban water conservancy system that combines "guidance and storage" and developed building waterproofing technology that adapts to multi-water environment. It has become a rare landscape garden city in northern China.

Living by the Water

The use of natural water sources is the natural ability of the ancients to survive. The ancient city of Jinan is dotted with springs both inside and outside, with abundant water sources, which attracted the ancients to settle and live here, and promoted production and economic and social development. The archaeological discovery of a large number of densely-distributed ancient wells in the ancient city, as well as the ditches for water diversion and the ancient roads built in response to the spring channels, are empirical evidence that ancients in Jinan adapted to the spring water environment and made full use of spring water resources. Living by the water is not only a natural geographical choice, but also shapes the cultural traditions and life wisdom of Jinan.

Wells, Canals and Roads of the Historic Site of Gaodusi Alley

Historic Site of Gaodusi Alley is located near the West Gate of the ancient city. They date from the Spring and Autumn Warring States Period to the Ming and Qing dynasties. Archaeological excavations have revealed remains such as water surfaces, wells, kiln sites and cellars. More than 40 water wells were discovered within the historic site, and the wells were not far apart, most of them were two or three meters away. It is very rare to have such a large number of wells distributed in a limited area. It is conceivable that as early as more than 2,000 years ago, this area was a place with strong water sources and dense population.

水井与地下水位

从井的深度看，战国水井一般深 2—3 米，井下便是含水的生土层，而晚期的水井底部仅停留在早期的文化堆积上，说明济南的地下水位是随着年代的推移、地层的堆积而不断提高的。

Wells and Groundwater Levels

Seen From the depth of the wells, in the Warring States period, water wells were generally 2-3m in depth, and the underground was the soil layer containing water. In the late period, the bottom of the wells only stayed on the early cultural accumulation, indicating that the groundwater level in Jinan continuously increased with the passage of time and the accumulation of strata.

发掘现场发现多口古井
Multiple Ancient Wells Discovered at the Excavation Site

宋代水井立面
Elevation of the Wells of the Song Dynasty

宋代水井立面
Elevation of the Wells of the Song Dynasty

陶井圈

Pottery Walling Crib

战国（前 476—前 221）
高都司巷遗址出土
济南市考古研究院藏
直径 80 厘米，高 90 厘米

泥质灰陶，胎土细腻，形体规整。

宋代泉水街巷

在高都司巷以南，西临护城河的区域发现有 3 条南北向、1 条东西向建设在砖砌沟渠边的道路。路边设置有导流槽，途经居民皆可直接取用槽内源源不断流淌的泉水。这表明，济南人民在宋代便有了依傍泉渠建设街巷的传统。

Spring Water Streets and Alleys in the Song Dynasty

In the area to the south of Gaodusi Alley and to the west of the moat, three north-south and one east-west roads were discovered, built along brick ditches. There was a diversion channel on the side of the road, and passing residents could directly access the continuously flowing spring water in the channel. That is, in the Song Dynasty, the people of Jinan had a tradition of building streets and alleys in accordance with the spring canal.

“W”纹灰陶罐

Gray Clay Pot with “W Patterns”

战国（前 476—前 221）
高都司巷遗址出土
济南市考古研究院藏
高 22.5 厘米，口径 16 厘米，底径 6 厘米

敞口，方唇，高领，折肩，弧腹，小平底，肩部刻划“W”纹，腹部饰绳纹。

戳印纹灰陶罐

Gray Clay Pot

战国（前 476—前 221）
高都司巷遗址出土
济南市考古研究院藏
高 27.5 厘米，腹径 26.3 厘米

敞口，方唇，高领，折肩，斜弧腹内收，小平底。肩饰多道弦纹，下腹近底处有一戳印纹。

白釉瓷瓶

White-glazed Porcelain Bottle

唐（618—907）
高都司巷遗址出土
济南市考古研究院藏
高 18.2 厘米，腹径 9.5 厘米

敞口外撇，圆唇，颈较长，鼓肩，弧腹逐渐内收，喇叭状高圈足。通体施白釉，圈足施青釉。

白釉瓷罐

White-glazed Porcelain Pot

元（1271—1368）
高都司巷遗址出土
济南市考古研究院藏
高 7.4 厘米，腹径 11.6 厘米

直口，圆唇，短颈，折肩，鼓腹，平底圈足。红褐胎，整体施白釉，遍布细碎开片，釉不及足。

酱釉瓷罐

Brown-glazed Porcelain Pot

明（1368—1644）
高都司巷遗址出土
济南市考古研究院藏
高 19.4 厘米，腹径 19.5 厘米

卷沿，圆唇，短颈，溜肩，鼓腹，下腹内收，平底。通体施酱釉，釉不及足，露浅灰白色胎。

省府前街遗址的水井

省府前街遗址位于古城西部，考古人员在此发现水井、房址、灶址等遗迹，出土大量汉代至明清时期的陶瓷器，第一次清晰地清理出山陕会馆遗址。遗址内发现从战国至明清时期的水井 19 口。数量众多的水井及井内的生活容器证明，这一区域自战国至明清时期人类活动频繁，为人口密集区。

Water Wells at Historic Site of Shengfuqian Street of the Provincial Government

Historic Site of Shengfuqian Street of the Provincial Government is located in the western part of the ancient city. Remains such as water wells, house sites and stove relics have been discovered, and a large number of ceramics from the Han Dynasty to the Ming and Qing dynasties have been unearthed. For the first time, the historic site of the Shanxi-Shaanxi Guild Hall have clearly been cleaned up. 19 wells from the Warring States Period to the Ming and Qing dynasties were discovered in the site. The large number of water wells and living containers in the wells prove that this area has been a densely-populated area with frequent human activities from the Warring States Period to the Ming and Qing dynasties.

省府前街遗址发掘现场（图中的石头就是露出的房基和水井）
Excavation Site of the Historic Site of Shengfuqian Street(The stones in the picture are exposed foundations and ancient wells)

卫巷遗址的水井和水渠

古城南部的卫巷遗址共发现水井 29 眼，各时期水井形制各有不同，此外还发现了明代砖砌水渠，这充分显示了此区域自汉代以来人与水的完美融合。

Wells and Canals at the Historic Site of Weixiang

A total of 29 water wells were discovered at the historic site of Weixiang in the southern part of the ancient city. The shapes of water wells in each period were different. In addition, brick-built aqueducts in the Ming Dynasty were discovered, which fully demonstrated the perfect integration of man and water in this area since the Han Dynasty.

探方内发现多口古井
Multiple Ancient Wells Discovered in the Excavation Trench

发掘中的汉代水井
Water Wells of the Han Dynasty under Excavation

石砌唐代水井
An Ancient Stone-Made Well of the Tang Dynasty

砖砌宋代水井
Ancient Stone-Made Wells of the Song Dynasty

明代水井及井边砖砌水渠
The Wells of the Ming Dynasty and Brick-Set Water Channel Beside

白釉“道德清静”瓷瓶

White-glazed Porcelain Vase Inscribed “Dao De Qing Jing” (Morality and Purity)

元（1271—1368）
卫巷遗址出土
济南市考古研究院藏
高 29.9 厘米，腹径 21.4 厘米

小口，溜肩，橄榄球鼓腹，圈足。肩颈部附加 4 个弦纹立耳。外施白釉至中腹部，内绘弦纹和黑花。中腹部以下施黑釉。肩部一周黑釉草书“道德清净”。

酱釉双鱼瓷瓶

Brown-glazed Porcelain Vase with Twin Fish Pattern

唐（618—907）
卫巷遗址出土
济南市考古研究院藏
高 27.9 厘米，口径 6.4 厘米

造型以鲤鱼为原型，瓶身为两条腹部紧贴、呈对称状的鱼形。鱼头部四目突起，鱼身刻划鱼鳞，上部前后各贴胸鳍一对，下部各贴腹鳍一对，两侧贴塑背鳍，中空为系。通体施酱釉。

小明湖遗址的沟渠和道路

小明湖遗址位于大明湖西南门对面，西侧紧邻护城河。考古发掘清理了 4 条宋代沟渠和 1 条沿沟渠而建的唐宋道路。在宋代地面上，相距 10 米之内即有 3 条南北走向的沟渠，证明本区域在宋代时期即有泉水丰沛的景象。砖砌沟渠系人工水利工程，其作用是把大沟渠内的水引至所需之处。

The Ditches And Roads of Xiaoming Lake Site

Historic site of Xiaoming Lake is located opposite the southwest gate of Daming Lake, and the west side is adjacent to the moat. Archaeological excavations cleared four ditches of the Song Dynasty and one road along the ditch of the Tang-Song period. There are three north-south ditches within 10m of the ground of the Song Dynasty, which proves that the area was rich in spring water during the Song Dynasty. Brick-built ditches are artificial water conservancy projects, functions of which are to divert water from large ditches to the desired location.

小明湖遗址发掘全景
Excavation Panorama of the Historic Site of the Xiaoming Lake

宋代沟渠

其中3条为南北向，1条为东西向。南北向沟渠深、宽均为1米。东西向沟渠长3.6米、高0.2米、宽0.5米，由3层砖砌筑，底部用砖平铺，与南北向沟渠垂直，由西向东引水至南北向沟渠中。

Ditches of the Song Dynasty

Of these ditches, 3 ditches are in the north-south direction and one ditch is in the east-west direction. The depth and width of the north-south ditches and canals are 1m. The east-west ditch is 3.6m in length, 0.2m in height, and 0.5m in width. It is built with three layers of bricks, and the bottom is laid flat with bricks, perpendicular to the north-south ditch. Water is drawn from west to east into the north-south ditch.

小明湖遗址发掘出的水沟
Ditch Excavated in the Historic Site of the Xiaoming Lake

唐宋道路

小明湖遗址内发现唐宋时期道路 1 条，为南北走向，宽 2.1 米，中间发现两条平行相距 0.5 米的古代车辙，为交通工具长期碾压形成。道路与沟渠平行，可以断定该道路就是水面旁边的主要交通通道。

Road of Tang and Song Dynasties

A road from the Tang and Song dynasties was discovered at the historic site of Xiaoming Lake, which runs north-south and is 2.1m in width. Two ruts parallel to each other and separated by 0.5m were discovered in the middle, which were formed by long-term crushing of transportation vehicles. The road runs parallel to the ditch, indicating that it is the main transportation channel next to the water surface.

小明湖遗址发掘出的车辙
Ruts Excavated in the Historic Site of the Xiaoming Lake

黑釉“西”字弦纹瓷瓶

Black-glazed Porcelain Vase with “Xi” Inscription and String Pattern

明（1368—1644）
县西巷遗址出土
济南市考古研究院藏
高 14.8 厘米，腹径 17.5 厘米

器身施黑釉，施釉至下腹部，釉色黑亮。颈部下方饰有弦纹两周，下书“西”字，腹部饰有弦纹 3 周。

白底褐彩“宜春奈夏”瓷枕

White-glazed Porcelain Pillow

明（1368—1644）
县西巷遗址出土
济南市考古研究院藏
长 26 厘米，最高 11.5 厘米，最宽 19 厘米

生活用具，为明代典型腰圆形制，枕面前低后高，通体施白釉。以褐彩为饰，枕面行书题写“宜春奈夏”四字，用笔厚重，字形舒展，既寄托对春秋顺遂的祝祷，又凸显消夏安眠的实用价值。集实用性与艺术性于一体，在磁州窑同类瓷枕中堪称匠心独运之作。

青釉刻花瓷盘

Celadon-glazed Porcelain Plate with Carved Floral Decoration

元（1271—1368）
县西巷遗址出土
济南市考古研究院藏
口径 15.1 厘米，底径 8.6 厘米，高 4.4 厘米

圆唇，敞口，浅腹，圈足。器内外施青釉，盘内刻有花卉纹。

青釉八思巴文瓷碗

Celadon-glazed Porcelain Bowl with Phags-pa Script Inscription

元（1271—1368）
县西巷遗址出土
济南市考古研究院藏
高 7.8 厘米，口径 19.9 厘米，足径 6.5 厘米

侈口，深曲腹，圈足。胎灰白，器壁内外施青釉。圈足内中心有墨书八思巴文。

白底黑花瓷罐

White-background Porcelain Jar with Black Flowers

元（1271—1368）
运署街遗址出土
济南市考古研究院藏
高 18 厘米，口径 15.5 厘米，足径 8.7 厘米

通体施白釉至底，领、肩、鼓腹处饰有多条黑带纹，上腹部绘有两组写意黑花纹。

钧釉瓷碗

Jun-glazed Porcelain Bowl

元（1271—1368）
运署街遗址出土
济南市考古研究院藏
口径 15.8 厘米，足径 5.6 厘米

天蓝色釉，釉厚，口沿处呈深褐色，器内满釉，外施釉不及足，露灰白胎。

白釉瓷执壶

White-glazed Porcelain Holding Pot

唐（618—907）
按察司街遗址出土
济南市考古研究院藏
高 15.5 厘米，腹径 15.5 厘米

侈口，束颈，溜肩，垂鼓腹，饼状实足。短直流，兽首形把手。

城市营造

济南古城在营造过程中，基于对泉水水量、分布特点、出露特点、汇流特点等考量，在城市选址、规模、布局及城垣形制等方面形成因借泉水而又兼容中国传统营造观的城市空间形态特征。济南古城选址结合传统“得水为上”的风水观，在独立的泉水系统下，以珍珠泉为穴；以趵突泉、黑虎泉环萦；以南依千佛山、北面黄河的态势形成以泉为脉的枕山、环水、面屏的城市选址格局。

Urban Creation

In the construction process of the ancient city of Jinan, based on the considerations of spring water volume, distribution characteristics, outcrop characteristics, confluence characteristics and other factors, the urban space morphology characteristics that are compatible with the traditional Chinese concept of construction due to the use of spring water have been formed in terms of urban site selection, scale, layout and city wall shape. The site selection of the ancient city of Jinan combines the traditional geomantic conception of “water is the top priority”. Under an independent spring water system, Pearl Spring is used as the cave; Baotu Spring and Black Tiger Spring are used as the surroundings; the situation of Qianfo Mountain to the south and the Yellow River to the north forms a pattern of urban site selection with springs as the veins of pillow mountains, surrounding water, and facing screens.

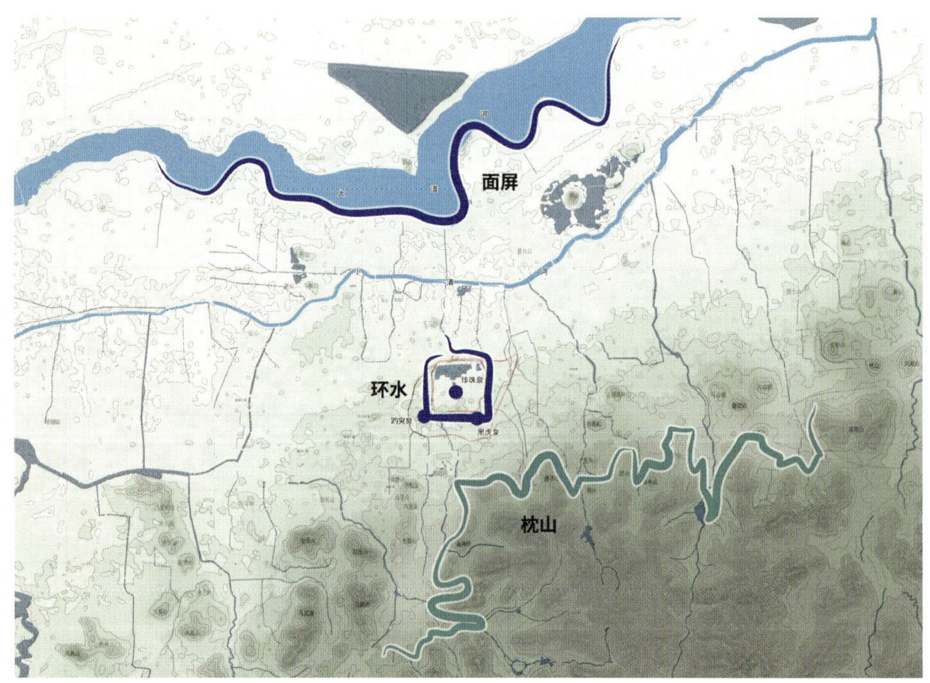

以泉为脉的枕山、环水、面屏的城市选址分析图
Analysis Diagram of Mountain-Backed, Water-Encircled City Siting with Spring-Fed Water Network and Fronting a Natural Screen

因泉随形

济南古城作为府城，其形制在受到传统平原城市“四方为形”的礼制规整布局影响的同时，因地制宜，城垣边界在基本方正的基础上，又营建出城墙四隅转角，以顺应水势，抗击雨洪。古城纳入调蓄泉水的约合三里的泉湖——大明湖，形成遵循府城“九里”规制，同时又能有效调蓄泉水的“十二里”的同时期府城中较大的城周规模。

Urban Layout Based on the Spring Water

As a prefectural city, the shape and structure of ancient city of Jinan is influenced by the traditional plain city's “square-shaped” ritual system and regular layout. Meanwhile, according to local conditions, the boundaries of the city walls are basically square. On the basis of the foundation, the four corners of the city wall have been built to conform to the water trend and fight against rain and floods. The ancient city integrated Daming Lake, a spring-fed reservoir spanning approximately three li(≈1.5 km), into its water regulation system. This created a perimeter of “twelve li” (≈6 km), which aligned with the traditional “nine-li” (≈4.5 km) framework for prefectural capitals while enhancing spring-water management, establishing it as one of the larger cities of its era in terms of hydraulic and spatial scale.

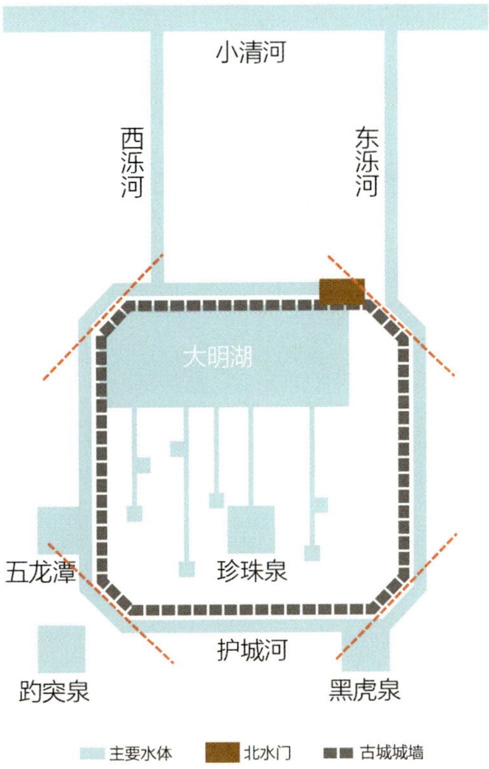

济南古城城墙转角示意图
Schematic Diagram of Corner Treatment for the City Wall of Ancient City of Jinan

流量较大的泉水均位于古城之外，城墙与护城河便成为阻止大量泉水入城的主要防御设施。流量较大的趵突泉和黑虎泉位于城墙的西南角及东南角，泉水的不断冲击增大了城墙转角处的压力。因此，济南古城城墙采用折线转角的形式，疏解了外围泉水对城墙的直接冲击。

济南城墙一角
A corner of City Wall in Jinan

玉带环城

济南护城河是国内唯一河水全部由泉水汇流而成的护城河，全长6.9公里，河道宽10—30米，最终流入大明湖。历史上，济南护城河具有重要战略防御功能。其河道经小清河通海，运输也是它的一大功用。此外，护城河还能够排涝、乘船观赏。

A city Moat Surrounding the City

The moat of Jinan is the only moat in China where all river water is formed by the convergence of spring water. It is 6.9km in length and 10-30m in width, ultimately flowing into Daming Lake. Historically, the Moat of Jinan had an important strategic defense function. Its river runs through the Xiaoqing River and connects to the sea, and transportation is also one of its major functions. Furthermore, the moat served dual purposes, including storm-water drainage (to mitigate flooding) and leisure activities such as boat tours for scenic viewing.

南护城河黑虎泉段
The Section of Black Tiger Spring of the South City Moat

四门不对

济南古城没有传统平原城市四门正对的十字街格局，城门位置结合传统礼制对位思想和冷泉分布及地势汇流的特点，形成西门偏南临近趵突泉、南门偏东避让名泉古鉴泉同时对位千佛山、北门偏东位于泉水汇流处、东门偏北对位齐国都城的“四门不对”特征。

Non-aligned Four-gate Configuration

In contrast to the traditional cross-axis street grid of flatland cities (with four gates aligned symmetrically), the ancient city of Jinan developed a “non-aligned four-gate configuration”. This layout harmonized classical ritual principles with hydrological and topographic realities. Specifically, the distribution of cold springs and drainage convergence. Key features include West Gate(adjacent to the Baotu Spring complex), South Gate (Shifted southeast to avoid the historic Gujian Spring while aligning symbolically with Qianfo Mountain (Thousand Buddha Mountain), North Gate (Located northeast at the confluence of spring-water channels), and East Gate (Set northwestward to ritually correspond with the ancient Qi State’s capital orientation).

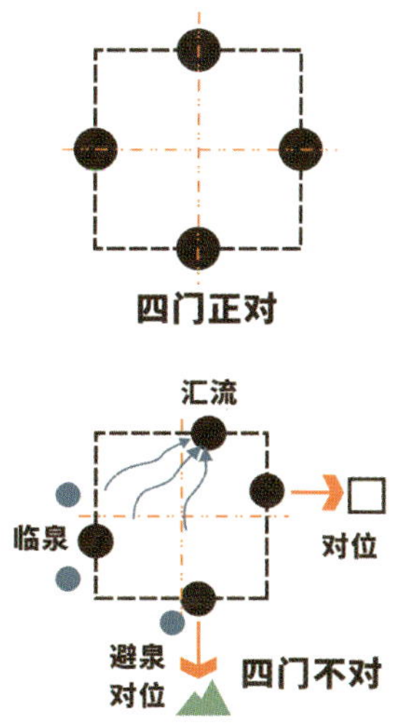

城门位置图
Schematic Diagram of the City Gate Position

城门位置关系示意图
Schematic Diagram of the Positional Relationship of City Gates

东门：齐川门

济南古城在先秦时期是“济右走廊”的重要节点，重要的“三齐门户”。东门的位置延续魏晋时期郡城肌理，连接东门大街一线，体现对位东侧的齐国临淄故城的礼制关系。

西门：泺源门

西城墙外趵突泉区域自宋代以来便成为重要的码头口岸，联系小清河航道，趵突泉也是小清河的主要水源。随着航运的发展，通航规模的扩大，五龙潭泉群周边也形成另一处港口。西侧的出入口成为水陆转换、城内城外商品输入输出的重要节点，因而在两泉群之间，即西城墙偏南处设西门。

East Gate: Qichuan Gate

In the pre-Qin period, the ancient city of Jinan served as a pivotal node along the “Jiyou Corridor” and a critical “gateway to the Three Qi” (historical term denoting Jinan’s role as the administrative and military threshold to the three core regions of the Qi State). The East Gate’s location perpetuated the urban fabric of the commandery city from the Wei and Jin dynasties period, aligning with the axis of East Gate Street. This design embodied a ritual-spatial alignment with the ancient Qi capital of Linzi to the east, reinforcing Jinan’s symbolic connection to its political predecessor.

West Gate: Luoyuan Gate

The Baotuquan Area outside the West City Wall has become an important wharf port since the Song Dynasty, connecting the Xiaoqing River channel, and Baotuquan is also the main water source of the Xiaoqing River. With the development of shipping and the expansion of navigation scale, another port has also formed around the spring group of Five Dragons Pool. The entrance and exit on the west side has become an important node for the conversion of water and land, as well as the import and export of goods inside and outside the city. Therefore, the west gate was set up between the two spring groups, located south of the west city wall.

木雕力士像建筑饰件

Wood-carved Architectural Ornament of Guardian Warrior Statue

元（1271—1368）
济南市博物馆藏
长 26.1 厘米，宽 18.4 厘米，高 8.9 厘米
由整块金丝楠木雕刻而成，明代重修济南城时被用于泺源门箭楼之上。

1928 年五三惨案后，力士像木构件被带去日本。1956 年，中日互派代表团访问，日方将这件力士像归还中国。

南门：舜田门（历山门）

济南古城南侧约 2.5 公里处为千佛山，同时天地坛街轴线沿线，南城墙外为名泉古鉴泉。为避让名泉古鉴泉，同时又基于礼制的山川崇拜对位千佛山，而在南城墙略偏东处设南门。

South Gate: Shuntian Gate (Lishan Gate)

Approximately 2.5km south of Jinan's ancient city lies Qianfo Mountain, situated along the axis of Tiantan Street, with the renowned Gujian Spring located outside the southern city wall. In order to avoid the famous ancient springs and the worship of mountains and rivers based on ritual, the south gate was built slightly east of the south city wall to align with Qianfo Mountain.

石兽

Stone Beast

元（1271—1368）
南门桥出土
济南市博物馆藏
长 66 厘米，宽 29 厘米，高 101 厘米

石兽头顶一角，双目圆睁，方口呲牙，身披鳞片，前腿直立，后腿蹲坐，背倚石座，整体造型古朴。

北门：汇波门（会波门）

济南古城北侧低洼之地被泉水汇集的大明湖占据，结合对泉水及雨洪的调蓄功能，在地势最低处，即北城墙偏东处设北水门。

North Gate: Huibo Gate

The low-lying northern area of ancient city of Jinan was occupied by Daming Lake, a basin fed by converging springs. To manage the dual functions of spring water regulation and flood control, the North Water Gate was constructed at the lowest topographic point, located along the eastern section of the northern city wall.

北水门历史影像
Historical Images and Status Quo of North Water Gate

泉水蓄导

济南古城的泉水蓄导系统改善了无序出露、宣泄的泉水环境，既保证了居民用水，又不至造成城内水患。古城的泉水储蓄主要是将大明湖加以改造加固，作为主要调蓄设施，同时利用其它小型自然水体作为补充。泉水疏导主要是通过疏通自然河道、建设人工水道等，一方面使泉水迅速北流，保障城内免受城外大泉威胁；另一方面迅速地排出城内多余泉水，使城区免受水害。

Storage and Diversion of Spring Water

Spring water storage and diversion system in the ancient city of Jinan improved the disordered exposed and cathartic spring water environment, which not only ensured the water use of residents, but also didn't cause flooding in the city. The spring water storage in the ancient city mainly involves the renovation and reinforcement of Daming Lake as the main storage facility, while utilizing other small natural water bodies as supplements. The diversion of spring water is mainly through the dredging of natural rivers and the construction of artificial waterways. On the one hand, it enables the spring water to quickly flow northward, ensuring that the city is not threatened by large springs outside the city; on the other hand, the excess spring water in the city is quickly discharged to protect the city from water damage.

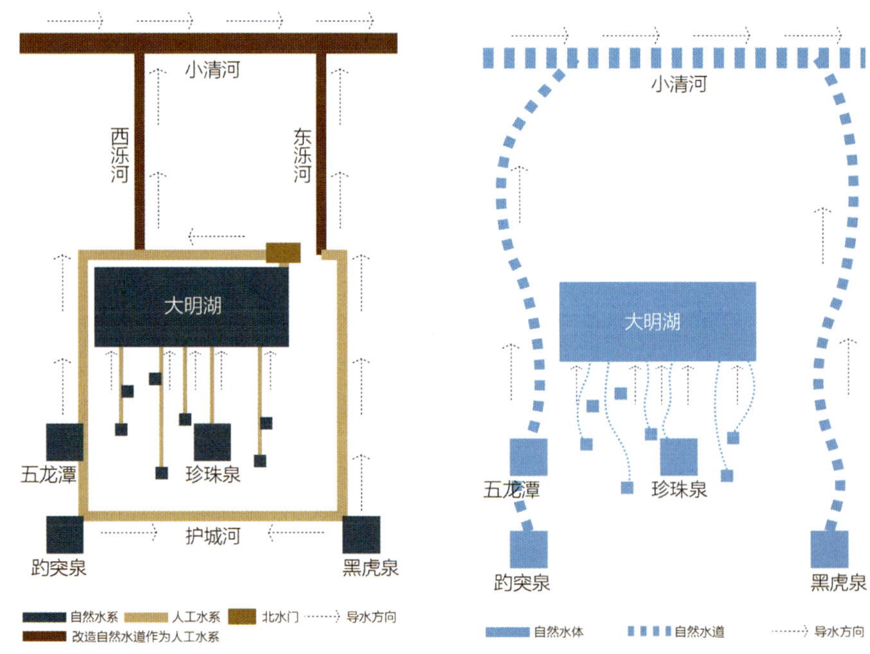

济南古城泉水蓄导示意图
Schematic Diagram of Storage and Diversion of Spring Water in the Ancient City of Jinan

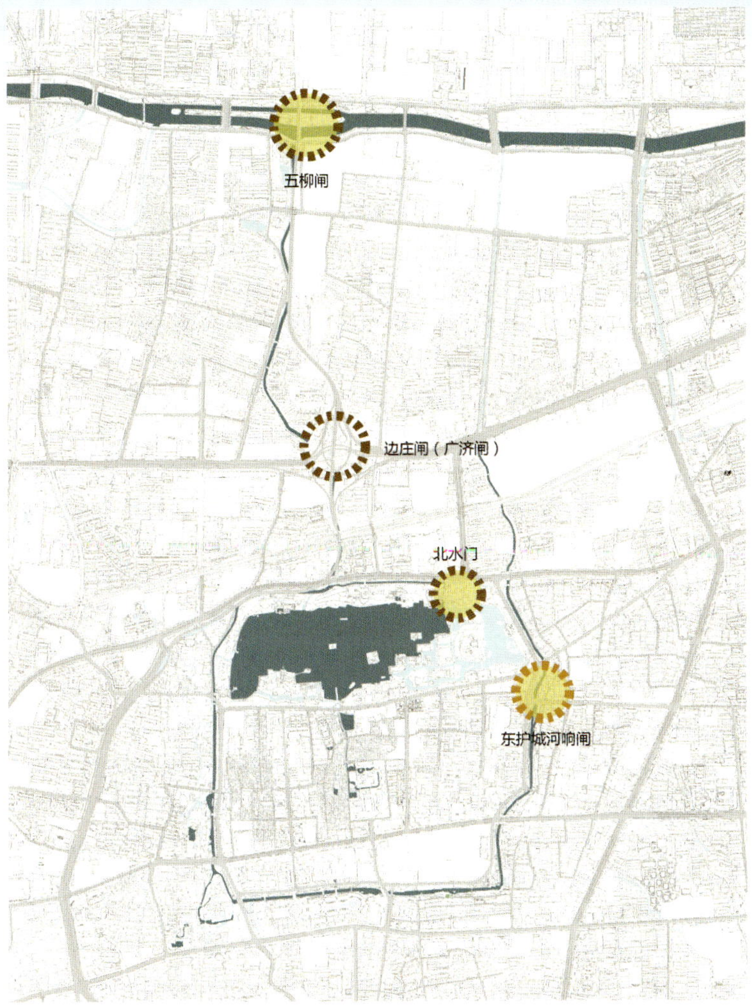

水闸分布图
Distribution Diagram of Sluices

历代济南人民建设了各式水闸，对泉水导蓄进行人工干预，使其在时间和空间上重新分配。泉水在古城区域由南向北排泄，在经过不同水闸时，闸门可根据水量调节开闭程度，从而起到阻挡水流和控制水量的作用。

清代济南府城街巷格局
The Pattern of Streets and Alleys in Jinan Prefecture during the Qing Dynasty

济南古城街巷除交通功能外，也可以作为泉水疏导设施。通常情况泉水出露水量较小，可通过泉渠疏导，水量较大时则会溢出，通过街巷疏导。古城南北向街道多且长，易于积水排泄；东西向街道少且短，用于引导积水。现存芙蓉街、曲水亭街都是典型的可用于泉水疏导的街巷。

独特的木桩基础建筑

由于长期生活在多水环境中，济南古城发展出了柏木桩基础、油泥坯墙面等多种建筑防水技术。木桩基础建筑主要分布在古城东北部、西北部和中西部。这是由于古城北半部的地下水位较高，土层含水量大，地基软弱，荷载力很低，易发生沉陷，毁坏建筑。而木桩基础是一种古老的地基处理方式，根据不同部位土层强度的差异，打入地下的木桩或穿透软弱土层支撑在坚实的土层中，或横向挤压周围土层以增加地基强度。同时，利用木桩与土层间的摩擦力，可以减少软弱地基对建筑的破坏。

Unique Buildings of Wooden Pile Foundations

Due to long-term living in water abundance environment, various building waterproofing technologies such as cypress pile foundations and oil-mud brick walls were developed in ancient city of Jinan. Buildings of wooden pile foundations were mainly distributed in the northeast, northwest and central western parts of the ancient city. This is because that northern half part of the ancient city of Jinan experiences notably elevated groundwater levels, leading to high soil moisture content, weak foundation bearing capacity, and susceptibility to subsidence and structural damage. The wooden pile foundation is an ancient method of foundation treatment. According to thc differences in the strength of the soil layers in different sites, wooden piles are driven underground or penetrate weak soil layers to support solid soil layers, or squeeze the surrounding soil layer laterally to increase the strength of the foundation. Meanwhile, utilization of the friction between wooden piles and soil layers can reduce the damage of weak foundations to buildings.

县西巷遗址明清时期大型建筑F7室内墙下木桩基础
Wooden Pile Foundation under the Interior Wall of F7 (a Large Building during the period of Ming and Qing dynasties) at the Historical Site of Xianxi Alley

趵突泉北路 6 号遗址城墙东侧清代房址木桩基础
Wooden Pile Foundation of Qing Dynasty House Site on the East Side of the City Wall at Historic Site on No. 6 North Baotu Spring Road

省府前街遗址东区明代木桩基础
Wooden Pile Foundations of the Ming Dynasty in the Eastern District at the Historic Site of Shengfuqian Street

大明湖综合整治工程发掘现场发现的清代木桩基础
Pile Foundations of the Qing Dynasty Discovered on the Excavation Site of the Daming Lake Comprehensive Improvement Project

福慧禅林前殿柱础下的基础木桩
Wooden Pile Foundations under the Column Bases of the Front Hall of the Fuhui Buddhist Temple

陶瓦当

Eaves Tile

汉（前 202—220）
东平陵城遗址出土
章丘区博物馆藏

瓦当，俗称“瓦头”，位于瓦的顶端，既能保护屋檐下椽头不受风雨侵蚀，又具有装饰的作用。瓦当最早出现于西周，一直沿用至明清，因其上装饰各色纹样而成为古代重要艺术品，并为世人珍藏。

兽面纹瓦当

Eaves Tile with Beast Face Patterns

宋（960—1276）
旧军门巷遗址出土
济南市考古研究院藏

圆形，当面为兽面浮雕。

莲花纹瓦当

Eaves Tile with Lotus Patterns

宋（960—1276）
十亩园宋代墓群出土
济南市考古研究院藏

圆形，当面为浮雕莲花纹饰。

龙纹瓦当

Eaves Tiles with Dragon Patterns

明（1368—1644）
宽厚所街遗址出土
济南市考古研究院藏

圆形，灰陶质，当面为龙纹浮雕。

砖雕建筑构件

Brick-carved Architectural Component

明（1368—1644）
济南市博物馆藏
长 32.5 厘米，宽 24 厘米，厚 10.5 厘米

构件雕一神兽，兽头前伸，双目圆睁，作张嘴大吼状，鬣（liè）毛长飘于脑后，神态威武。

绿琉璃龙首吻

Green-glazed Dragon-head Roof Ridge Ornament

清（1644—1911）
县西巷福慧禅林遗址出土
济南市考古研究院藏
长 40 厘米，宽 15 厘米，高 37 厘米

吻为建筑构件，置于屋脊之上，既有稳定结构、防止漏水的实用功能，又兼具装饰作用。龙头昂首怒目，尾部向上向外卷曲，鳞飞爪张。有镇宅辟火之意。

地面提升

为了防止出现地下水位过高、遇雨宣泄不及的情况，济南古城多选择在地势较高处建府邸，并采取垫高地面或修筑台基的方式。此外，古城内也长期进行着提升地面工程，唐宋时期地面在现代地面约 5 米下。过去的房子大多是土坯房，废弃后就可直接推倒填地。明清时期许多衙门和府邸都在进行填地活动。

Ground Elevation

In order to prevent conditions where underground water is high and there is insufficient drainage in case of rain, the ancient city of Jinan often chose to build mansions at higher elevations and adopted methods such as raising the ground or building the stylobates. In addition, there had been long-term ground improvement projects in the ancient city, with the ground level being about 5m below the modern level during the Tang and Song dynasties. Most of the houses in the past were adobe houses, which could be directly bulldozed and reclaimed after they were abandoned. During the Ming and Qing dynasties, many yamens and mansions were engaged in land filling activities.

济南古城府邸用地抬高防水
Raised Waterproof Platform of the land for the Mansion of the Ancient City of Jinan

城市诗意

济南的泉与城之间存在着紧密的共生关系。分布广泛的泉水，与济南的自然和人文景观共同构成了山、泉、湖、河、城一体共生的宏大格局，塑造了独特的城市景观和生活方式。府城内外，依泉就景的泉水园林、泉景合一的人文景致，连缀成一幅典雅灵动的幽美画卷；古往今来，寄情泉林的名士追求、人泉融合的生活传统，奏响了一曲诗意栖居的隽永弦歌。济南城与泉水的共生关系，是自然与人文相互依存、相互促进的生动写照。

潇洒湖山

济南的泉多如繁星，风采绚烂。清澈的泉水是这座城城市的灵魂，翠湖青山、曲水潺缓，孕育出一代又一代名人贤士，也醉倒了无数的迁客骚人。他们通过诗词、歌赋、书画等形式，将泉水的美丽与自身的情感体验融为一体，创作出许多传世佳作，为济南这座城市增添了丰厚的文化内涵。千百年过去了，名士们已经在历史的沧桑风雨中远去，唯有清泉长流，吟唱着当年风流。

City Poetics

There is a close symbiotic relationship between springs and the city in Jinan. The widely distributed spring water, together with the natural and cultural landscapes of Jinan, constitute a grand pattern of mountains, springs, lakes, rivers and cities coexisting, shaping a unique urban landscape and way of life. Within and beyond the prefecture city, spring-nourished gardens harmoniously arranged around natural springs, and cultural landscapes seamlessly integrated with aquatic vistas, are woven into an elegant, living scroll of serene beauty. Throughout history, the pursuit of scholars who found solace in in spring-nurtured groves, as well as the tradition of integrating human and spring life, have composed a timeless symphony of poetic dwelling. The symbiotic relationship between Jinan City and springs is a vivid portrayal of the interdependence and mutual promotion between nature and culture.

Unfettered Elegance of Lakes and Mountains

The springs in Jinan are as numerous as stars, with brilliant charm. The clear spring water is the soul of this city, with lush green mountains and winding waters that have nurtured generations of celebrities and scholars, as well as intoxicated countless migrant workers. They integrate the beauty of spring water with their own emotional experiences through poetry, songs, calligraphy and other forms, creating many timeless masterpieces and adding rich cultural connotations to the city of Jinan. Thousands of years have passed, and the famous scholars and celebrities have gone far away in the vicissitudes of history, but the clear springs are flowing, singing about the past romance.

紫檀木雕竹节形臂搁

Rosewood-Carved Bamboo-Joint-Shaped Armrest

清（1644—1911）
济南市博物馆藏
长 18.2 厘米，宽 7.2 厘米，厚 3.8 厘米

整体雕成竹节形。臂搁，是中国古代文人写字作画时用来搁放手臂的文案用具。

高凤翰铭天鹅随形端砚

Natural Form-retained Duan Inkstone

清（1644—1911）
济南市博物馆藏
长 20.6 厘米，宽 13.4 厘米，厚 2.3 厘米

随形，巧妙地雕琢修饰为一只若隐若现的天鹅。

朱文震铭黄云紫电随形端砚

Natural Form-retained Duan Inkstone

清（1644—1911）
济南市博物馆藏
长 16.4 厘米，宽 14.1 厘米，厚 3 厘米

随形。石质细腻滑润，周围有一圈黄色晕纹，恰似黄云环绕。

莲蓬玉洗

Lotus Seedpod-Form Jade Writing-brush Washer

明（1368—1644）
济南市博物馆藏
长 16.2 厘米，高 8.3 厘米

玉质白润。圆雕洗子呈莲蓬状，口呈“如意云”状，莲蓬子凸起饱满。底部雕一荷叶将莲蓬托起。

竹雕人物故事笔筒

Pen Holder of Bamboo Carving Character Story

明（1368—1644）
济南市博物馆藏
高 13 厘米

竹质，直口，筒形，底缘系角器镶成。

汪节庵造书匣形墨

Book-Case-Shaped Ink Cake Made by Wang Jie'an

清乾隆三十年（1765）
济南市博物馆藏
长 8 厘米，宽 3.5 厘米，厚 1.3 厘米

墨为不规则书匣形。汪节庵，名宣礼，字蓉坞，安徽歙县人，歙派中的著名墨家。

听雪斋百蝠墨

Hundred-Bats Ink Cake from Tingxue Zhai Studio

清乾隆（1736—1796）
济南市博物馆藏
长 7 厘米，直径 1.5 厘米

墨为六棱圆柱形制，通体云纹衬地，遍饰漱金蝙蝠，其中一侧上端阴识楷书填蓝“听雪斋”，并嵌白色小珍珠一颗。

象牙雕花毛笔

Ivory Carved Writing-Brush

清（1644—1911）
济南市博物馆藏
长 22.1 厘米，直径 0.9 厘米

象牙质，通体雕刻花纹。

龙钮寿山长方印

Rectangular Seal

清（1644—1911）
济南市博物馆藏
长 2.6 厘米，宽 2 厘米，高 5.3 厘米

寿山石质。印章方形，雕一龙纽，印文阴文篆书“孔毓圻兰堂”。孔毓圻（1657—1723），号兰堂，山东曲阜人，孔子的第 67 代嫡长孙。

兽钮田黄长方印

Rectangular Seal

明（1368—1644）
济南市博物馆藏
长 2.8 厘米，宽 2.4 厘米，高 4.4 厘米

田黄石质。印章方形，质地温润。上雕一兽纽，印文阳文篆书“萝扶秋影到江乡”。

于阗（tián）采玉图玉山子

Yushanzi (Jade Ornament)

清乾隆（1736—1796）
济南市博物馆藏
长 20.1 厘米，宽 8.2 厘米，高 12 厘米

玉质青色，细腻光润。整体呈山形，采用圆雕、透雕、线刻等技法，巧妙地雕琢了于阗采玉工采得玉石的喜悦情景。此山子应为乾隆时期宫廷造办处的精品佳作，极其珍贵。

濮（pú）仲谦刻字竹节花插

Bamboo-joint Flower Arrangement

明（1368—1644）

济南市博物馆藏

高 12 厘米，底径 4.4 厘米

花插截取一段竹竿制成，旁边保留两支竹节，造型别致。在旁边小竹枝分别刻“右调百字令。乙酉三月”和“仲谦制”。濮仲谦，金陵派竹雕创始人。

三螭（chī）纹玉瑗（yuàn）

Jade Yuan (Large-Holed Annular Ring)

清（1644—1911）

济南市博物馆藏

直径 12.9 厘米，厚 1.8 厘米，高 9.3 厘米

玉质白润，光泽透明。玉瑷浮雕 3 条蟠螭，相互尾随呈环状。

双夔（kuí）耳寿桃玉瓶

Jade Bottle

明（1368—1644）
济南市博物馆藏
盖长 7.8 厘米，盖宽 3.9 厘米；瓶长 16.8 厘米，瓶宽 13.8 厘米；
通高 27.4 厘米

玉质青白。瓶盖顶部雕桃形纽，阴刻花瓣一周。瓶肩部镂雕夔龙形双耳，瓶两侧各雕一棵桃树自下而上盘绕瓶体，结仙桃多个，并有桃花盛开。

东坡游赤壁橄榄核雕

Nut-carving

清光绪（1875—1908）
济南市博物馆藏
长 4.2 厘米，宽 2.0 厘米，高 2.0 厘米

随橄榄核形雕华丽篷船，雕刻有多扇开启的窗户及精美纹饰。船头船尾各雕两人端坐茶几旁饮酒观景，船底精刻苏东坡《赤壁赋》7 行。

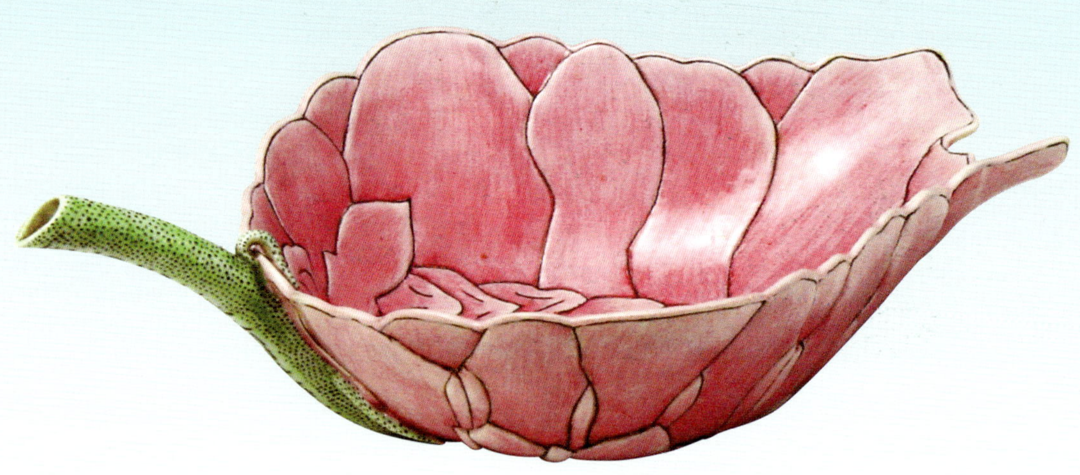

粉彩荷花式吸杯

Famille Rose Lotus Shaped Straw Cup

清（1644—1911 年）
山东博物馆藏

杯体作莲花瓣形，花柄中空可吸水，梗部以黑彩楷书“大清光绪三十四年安徽太湖附近秋操纪念杯”纪年款

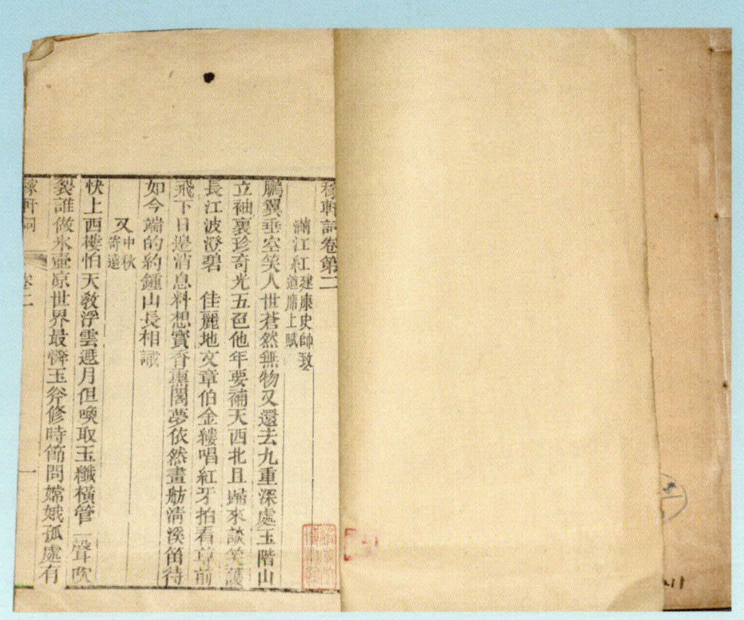

稼軒詞卷第二
滿江紅 建康史帥致道席上賦
鵬翼垂空笑人世蒼然無物又還去九重深處玉階山
立袖裏珍奇光五色他年要補天西北且歸來談笑護
長江波澄碧 佳麗地文章伯金縷唱紅牙拍看尊前
飛下日邊消息料想寶香黃閣夢依然畫舫清溪笛待
如今端的約鍾山長相識
又 中秋寄遠
快上西樓怕天教浮雲遮月但喚取玉纖橫管一聲吹
裂誰做冰壺涼世界最憐玉斧修時節問嫦娥孤處有

辛弃疾稼轩集抄存四卷补遗一卷刻本

Block-Printed Edition of Jia Xuan's Collected Manuscripts in Four Volumes of the Qing Dynasty with a Supplementary Volume

清（1644—1911）
济南市博物馆藏

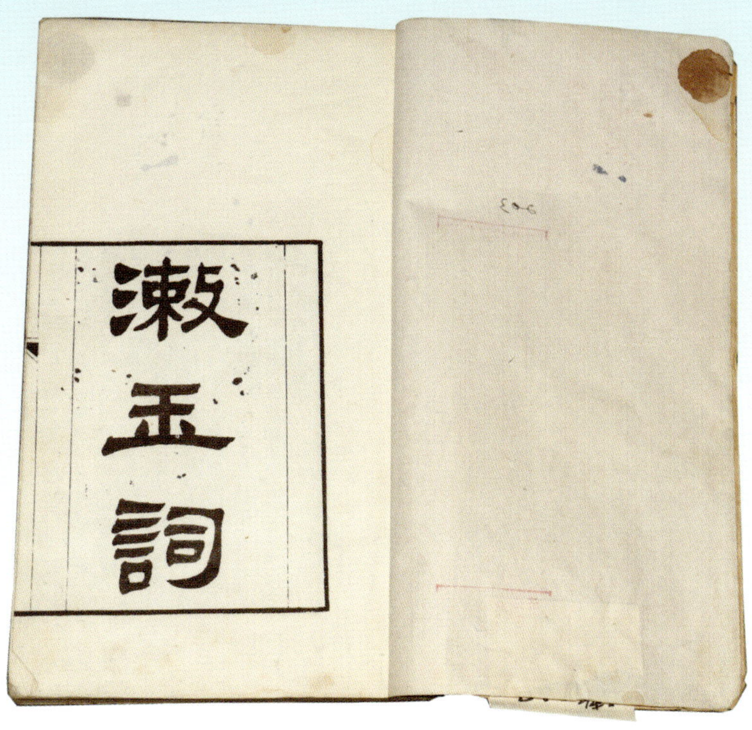

李清照漱玉词四印斋重刊本

Siyinzhai Reprinted Edition of Shu Yu Ci (Jade-Rinsing Lyrics) During the Reign of Emperor Guangxu of the Qing Dynasty

清光绪（1875—1908）
济南市博物馆藏

杜甫：济南名士多

唐天宝四年（745）夏，杜甫去临邑（今山东临邑县）看望任临邑主簿的弟弟杜颖，途经济南。当时恰逢济南（时称临淄郡）太守李之芳重建历下亭竣工，时任北海郡（治青州）太守李邕恰好也在济南，故友久别重逢，在历下亭宴集。名士相聚，觥筹交错，自然不能无诗，于是杜甫即席写下了“海右此亭古，济南名士多”的千古名句。

Du Fu: “Numerous Famous Literati in Jinan”

In the summer of the fourth year of Tianbao of the Tang Dynasty (745), Du Fu went to Linyi (now Linyi County, Shandong Province) to visit his younger brother Du Ying, who served as the the deputy governor of Linyi. He passed through Jinan. At that time, the reconstruction of Lixia Pavilion by Li Zhifang [the governor of Jinan (then known as Linzi Prefecture)] was completed. Li Yong, the prefecture chief of Beihai Prefecture (governing Qing Prefecture), happened to be in Jinan. After a long separation, the two old friends reunited and held a banquet at Lixia Pavilion. Amidst the gathering of famous literati and the clinking of glasses, it was inevitable that poetry could not be absent. Therefore, Du Fu immediately composed the immortal verses through the ages:“Ancient pavilion in Qi region,numerous famous literati in Jinan.”

杜甫（712—770），字子美，自号少陵野老，祖籍襄阳（今属湖北），生于巩县（今河南巩义），唐代著名现实主义诗人，后世尊称他为“诗圣”，称其诗为“诗史”。

Du Fu (712-770), courtesy name Zimei, self-styled as Shaoling Ye Lao (Old Hermit of Shaoling), was a renowned realist poet of the Tang Dynasty. Born in Gong County (Now Gongyi, Henan Province) with ancestral roots in Xiangyang (now Xiangyang, Hubei Province), he is posthumously honored as the “poet-sage”, and his works are celebrated as “history of poetry” for their vivid chronicling of societal upheavals during the mid-Tang period.

济南八景：历下秋风

如若秋日，微风吹来，历下亭风景更妙。不论是岸上的垂杨，还是水下的树影，皆舞动着青纱，婀娜多姿，好像神女在清波深院中起舞。湖里的荷花，在轻风中颤动，犹如江妃在水面上凌波微步，湖水晃漾，清波拍岸，如玉佩相击，发出清越的声响，如奏乐章。

历下亭位于大明湖东南隅岛上，因其南临历山而得名。历下亭是一座历史悠久、文化底蕴深厚的名亭。它始建于北魏，历经多次兴废变迁，至清初由山东盐运使李兴祖重建，成为济南的一大名胜。

Autumn Breezes of Lixia, One of the Eight Scenic Spots of Jinan

If the breeze blows in autumn, the scenery of Lixia Pavilion will be even better. Whether the weeping willows on the shore or the submerged shadows of trees beneath the water, all sway in their gauzy verdant veils, their lithe movements like a celestial maiden dancing in a courtyard embraced by clear, rippling waves. The lotus blossoms upon the lake tremble in the gentle breeze, like a river goddess pacing with ethereal grace atop the shimmering waves. The water ripples and sways, its crystalline currents lapping the shore with a clarity akin to jade pendants chiming in harmony, as though composing a celestial melody.

Lixia Pavilion is located on an island in the southeast corner of Daming Lake, named after its south facing Mount Li. Lixia Pavilion is a famous pavilion with a long history and profound cultural heritage. It was first built in the Northern Wei Dynasty and has gone through many ups and downs. In the early Qing Dynasty, it was rebuilt by Li Xingzu, the salt transport commissioner of Shandong, and became a major scenic spot in Jinan.

历下亭历史影像
Historical Images of Lixia Pavilion

曾巩：俯仰林泉绕舍清

北宋年间，曾巩曾任齐州知州。曾巩在济南任职虽只有两年，却对济南的名泉胜景情有独钟，留下了大量描写济南的诗文，其中不乏咏泉赞泉的名篇佳作。曾巩对大明湖进行整治，他主持修建的百花堤将大明湖分成西湖与东湖，被后人称为“曾堤”；还修建了百花台和北渚、环波、水香等亭，以及芙蓉、水西、湖西、北池、百花等七桥，使大明湖成为济南的一大名胜。

Zeng Gong: “Beneath my gaze, woods and springs encircle my dwelling in tranquil purity.”

During the Northern Song Dynasty, Zeng Gong served as chief of Qi Prefecture. Although Zeng Gong only served in Jinan for two years, he had a soft spot for Jinan's famous springs and scenery. He left a large number of poems describing Jinan, many of which are famous and excellent works praising springs. During his tenure as the governor of Jinan, Zeng Gong initiated comprehensive hydrological and landscape renovations at Daming Lake, transforming it into a cultural and ecological landmark. Baihua Causeway, which was constructed under his charge, divided Daming Lake into West Lake and East Lake, which was called “Zeng Causeway” by later generations. He also built Baihua Terrace and pavilions such as Beizhu, Huanbo and Shuixiang, as well as seven bridges such as Furong, Shuixi, Huxi, Beichi and Baihua, making Daming Lake a famous scenic spot in Jinan.

俯仰林泉绕舍清，
终年闲卧济南城。

曾巩（1019—1083），字子固，世称南丰先生，建昌军南丰（今属江西）人，北宋文学家、史学家、政治家，唐宋散文八大家之一。

Zeng Gong (1019-1083), courtesy name Zigu and revered as Master Nanfeng, was a native of Nanfeng, Jianchang Commandery (modern-day Nanfeng, Jiangxi Province). A polymath of the Northern Song Dynasty, he was distinguished as a literati, historian and statesman, and is enshrined among the Eight Great Prose Masters of the Tang and Song Dynasties, a canonical group of Chinese essayists.

济南八景：明湖泛舟

大明湖昔日湖面很大，碧波万顷，遥接群山，气势壮阔。在这宽阔的水域中，四季都呈现出不同的景色。春日，最为丰姿的要算是垂柳，湖岸、沙洲、岛上，无处不是，有的弯着，有的卧着，有的立着，清风一吹，柳丝轻荡。夏日荷浪迷人，片片的葱绿，点点的嫣红。深秋，雪白的芦花，谢落如柳絮，随风飘扬，翩翩起舞。冬日，明湖虽暂失绿波，但银装素裹也分外妖娆。明湖，不仅四季景色奇异，而且常年充满生机。黄鹂鸣于翠柳，鸥鹭点于清波，笙歌扬其远韵。人们便将这一景观称为“明湖泛舟”。

Rafting on Minghu Lake, One of Eight Scenic Spots of Jinan

The surface of Daming Lake used to be very large, with vast blue waves and distant mountains, exuding a magnificent atmosphere. In this vast water area, different scenery is presented throughout the four seasons. In spring, the most graceful ones are the weeping willows. Along the lakeshore, across sandbars, and upon the islands, willows thrive in every corner-arching, reclining, or standing tall. As a gentle breeze stirs, their slender branches sway in rhythmic waves, painting the air with whispers of grace. Enchanting under the sun, the lake ripples with swathes of emerald green, endless lotus leaves, punctuated by flecks of crimson, where blossoms blush like poetry on water. In late autumn, snow-white reed plumes shed like willow down, swirling and dancing on the wind's breath, a ballet of frost-touched elegance. In winter, though Ming Lake loses its emerald ripples, it dons a cloak of silver frost, a spectral beauty that haunts the ice-bound shores. The Ming Lake, not only does it unveil surreal beauty across four seasons, but it also hums with a timeless vitality. Golden orioles sing amidst emerald willows; egrets and herons grace crystal-clear ripples, as reed-pipe harmonies drift into distant echoes. As visitors go boating among them, they always feel “like wandering through a realm of fragrance”. People call this landscape as “Boating on the Ming Lake”.

大明湖历史影像
Historical Images of Daming Lake

苏辙：流泉海内无

宋神宗熙宁六年夏，时年 35 岁的苏辙由陈州（治所在今河南省淮阳县）学官改任齐州（治所在今山东省济南市）掌书记。此后，一直至熙宁九年十月任满离济，苏辙宦居济南 3 年多。他的许多诗文都与济南胜景有关，他也经常游览大明湖。济南秀美的湖光山色，给长期羁旅在外的苏辙带来莫大的心灵慰藉，冲淡了他心中那浓浓的乡愁。

应念兹园好，
流泉海内无。

Su Zhe: "No spring in all the realm flows as this spring -it is unrivaled under heaven."

In the summer of the sixth year of the Xining era under Emperor Shenzong of the Song Dynasty, Su Zhe, then 35 years old, was transferred from his post as Education Official of Chenzhou (administrative seat in present-day Huaiyang County, Henan Province) to serve as Chief Clerk of Qi Prefecture (administrative seat in present-day Jinan City, Shandong Province). From that time until October of the ninth year of the Xining era, when his official term concluded and he departed Jinan, Su Zhe resided and served in the city for over three years. Su Zhe's poetry and prose often celebrated the scenic wonders of Jinan. He once praised the city's springs with heartfelt admiration, He also frequently wandered the shores of Daming Lake, immersed in its tranquil vistas. The beautiful lake and mountain scenery of Jinan brought great spiritual comfort to Su Zhe, who had been away for a long time, and diluted the strong homesickness in his heart.

苏辙（1039—1112），字子由，一字同叔，晚号颍滨遗老，世称“苏文定公”。北宋文学家，“唐宋八大家”之一，文与父亲苏洵、兄长苏轼齐名，合称“三苏”。

Su Zhe (1039-1112), courtesy name Ziyou and Tongshu, later known by his literary sobriquet Yingbin Yilao ("Old Recluse of Yingbin"), was posthumously honored as Duke Wen of Su for his enduring contributions to literature and governance. A renowned Northern Song Dynasty literatus and one of "the Eight Great Prose Masters of Tang and Song Dynasties", Su Zhe stands alongside his father, Su Xun and his elder brother Su Shi, as part of the illustrious trio collectively celebrated as the "Three Sus".

李清照：常记溪亭日暮

相传宋代济南女词人李清照的故居在柳絮泉边，她幼时常到柳絮泉东北方的漱玉泉，对着清明如镜的泉水梳洗打扮，后来便把自己的词集命名为《漱玉集》。李清照的《如梦令》，描绘的极有可能就是她在济南游玩的美好时光。

Li Qingzhao: “Often remembering playing in the pavilion by the stream until the sun set.”

Legend has it that Li Qingzhao, the celebrated Song Dynasty female poet, once resided near Liuxu Spring in Ji’nan. As a child, she frequented Shuyu Spring northeast of Liuxu Spring, where she would groom herself by its mirror-clear waters. Inspired by this pristine spring, she later named her anthology of ci poetry as *Shuyu Collection*. Li Qingzhao’s *Like a Dream* is highly likely to depict her wonderful time playing in Jinan.

李清照（1084—1155），号易安居士，齐州章丘人，宋代婉约派代表词人。

Li Qingzhao (1084-1155), literary name Yi’an Jushi (Hermit of Easeful Dwelling), was a native of Zhangqiu, Qizhou (present-day Jinan, Shandong Province). She is celebrated as the foremost representative of the Wanyue School (“Graceful and Restrained School”) of ci poetry in the Song

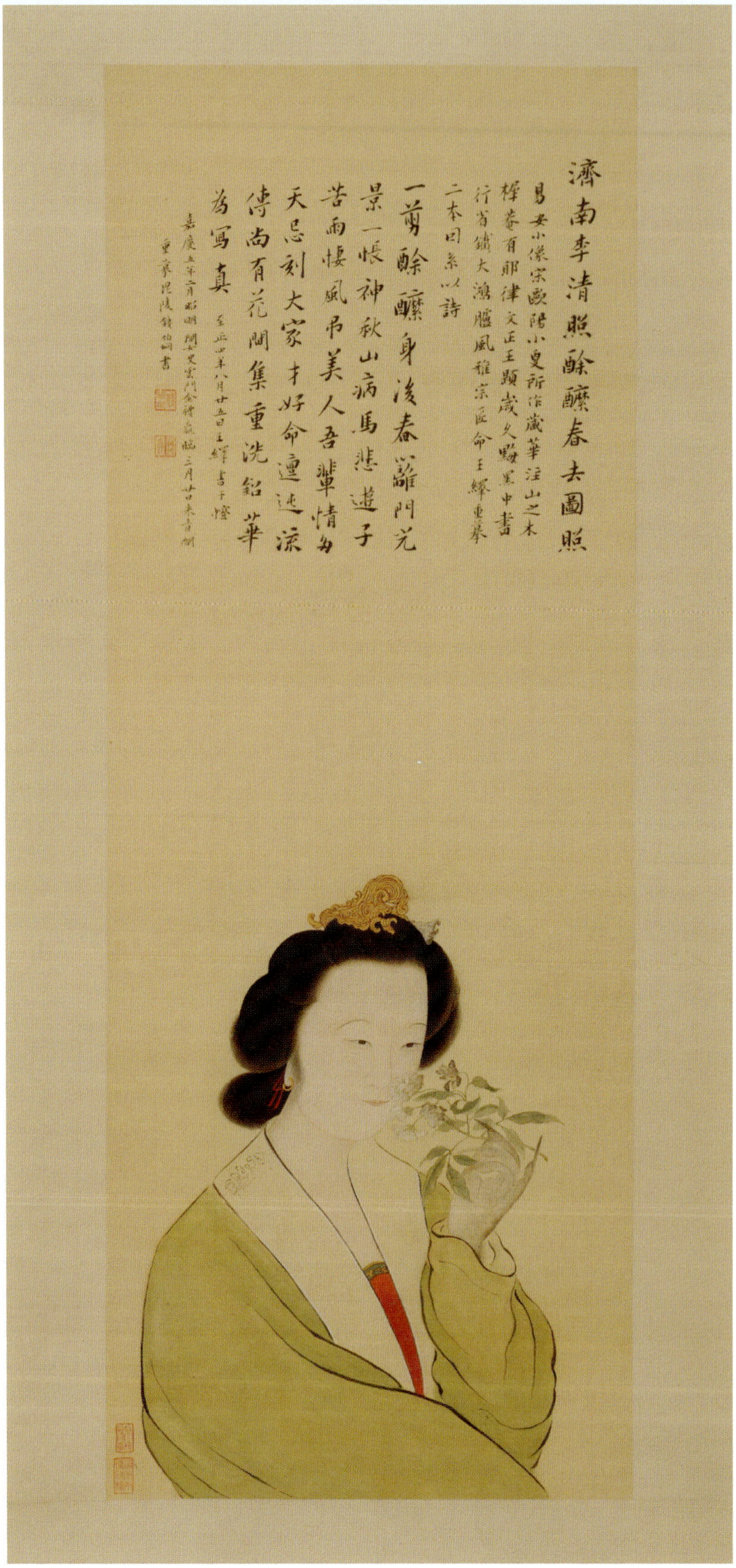

金礼嬴摹李清照酴(tú)醾(mí)春去图轴

Jin Liying's Facsimile Copy of Li Qingzhao

清（1644—1911）
济南市博物馆藏
长 105.2 厘米，宽 39 厘米

酴醾本是酒名，这里指花名。

赵孟頫：云雾润蒸华不注

济南趵突泉畔泺源堂的楹柱上，镶嵌着一副楹联："云雾润蒸华不注，波涛声震大明湖。" 它的作者就是元代著名书画家和诗人赵孟頫（fǔ）。赵孟頫在出任同知济南路总管府事期间，写下了许多歌咏济南湖光山色的优美诗篇，而其中最为著名的便是七律《趵突泉》。

Zhao Mengfu: Mist and Clouds Moisten and Rise over Hua Mountain

On the jacaranda column of Luoyuan Hall on the banks of Baotu Spring in Jinan, a couplet is inlaid: "Mist and clouds moisten and rise over Huabuzhu; Roaring waves resound through the Great Ming Lake." Its author is Zhao Mengfu, a famous calligrapher, painter and poet in the Yuan Dynasty. During his tenure as Vice Commissioner of Jinan Circuit's Administrative Office, Zhao Mengfu composed numerous poetic works celebrating the lakes and mountains of Jinan. Among these, the most renowned is his heptasyllabic regulated verse Ode to *Baotu Spring*.

云雾润蒸华不注，
波澜声震大明湖。

赵孟頫（1254—1322），字子昂，吴兴（今浙江湖州）人，宋末元初书法家、画家、文学家，同欧阳询、颜真卿、柳公权并称"楷书四大家"。他主张变革南宋画院的体制格调，开创了元代的新画风，为元代画坛的领袖人物，有"元人冠冕"之誉。

Zhao Mengfu (1254-1322), courtesy name Zi Ang, was a native of Wuxing (Now Huzhou, Zhejiang Province). A renowned calligrapher, painter, and literatus of the late Song and early Yuan Dynasties, he is celebrated as one of the "Four Masters of Regular Script", alongside Ouyang Xun, Yan Zhenqing and Liu Gongquan. He advocated for the reform of the system and style of the Southern Song Dynasty painting academy, creating a new style of painting in the Yuan Dynasty. He was a leading figure in the painting world of the Yuan Dynasty and was known as the "Crown of the Yuan Dynasty".

济南八景：鹊华烟雨

济南城北的鹊、华两山，山势俊秀，两山之间原为鹊山湖，湖光浩渺，碧波万顷，水村渔舍，柳绿花红，渚生蒲苇，水浮荷菱，微风轻浪，舟载歌声。阴云之际，两山连亘，若离若合，时隐时现，云雾缭绕，如二点青烟，诚为一幅绝妙的水乡图画，故昔人称此景为“鹊华烟雨”，为济南八景之一。

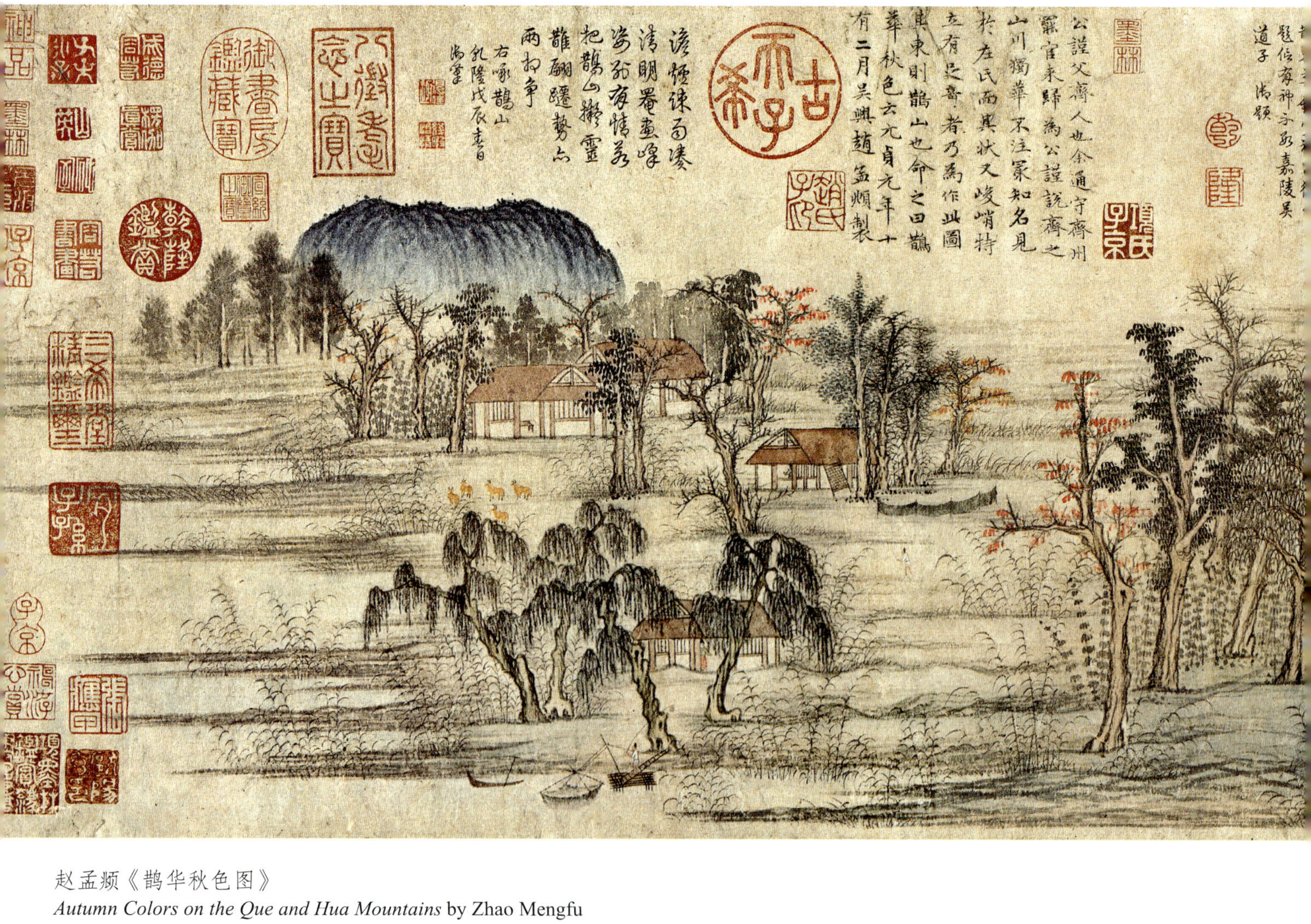

赵孟頫《鹊华秋色图》
Autumn Colors on the Que and Hua Mountains by Zhao Mengfu

Misty Rain over Qiao and Hua Mountains, One of Eight Scenic Spots of Jinan

North of Jinan City, the twin peaks of Que and Hua Mountains rise with elegant grandeur, their slopes adorned with lush greenery. Between them once lay Que Mountain Lake, a vast expanse of shimmering waters that stretched endlessly under the sky. Along its shores, waterside villages and fishing cottages nestled among willows cascading in emerald hues and blossoms ablaze in crimson. Reed marshes thrived on the islets, while lotus flowers and water chestnuts floated gracefully on the surface. Gentle breezes rippled the waves, and boats glided across the lake, their oars harmonizing with the melodies of fishermen's songs that echoed over the tranquil scene. Under brooding clouds, the twin peaks of Que and Hua Mountains stretch across the horizon, hovering between separation and union - now visible, now veiled, their forms shrouded in drifting mist. Like two ethereal wisps of blue-tinged smoke, they merge with the vaporous sky, composing an exquisite aquatic tableau. It is this poetic interplay of mountains, mist and rain that inspired the ancients to name the scene "Misty Rain over Que and Hua Mountains", considered as one of the Eight Views of Jinan.

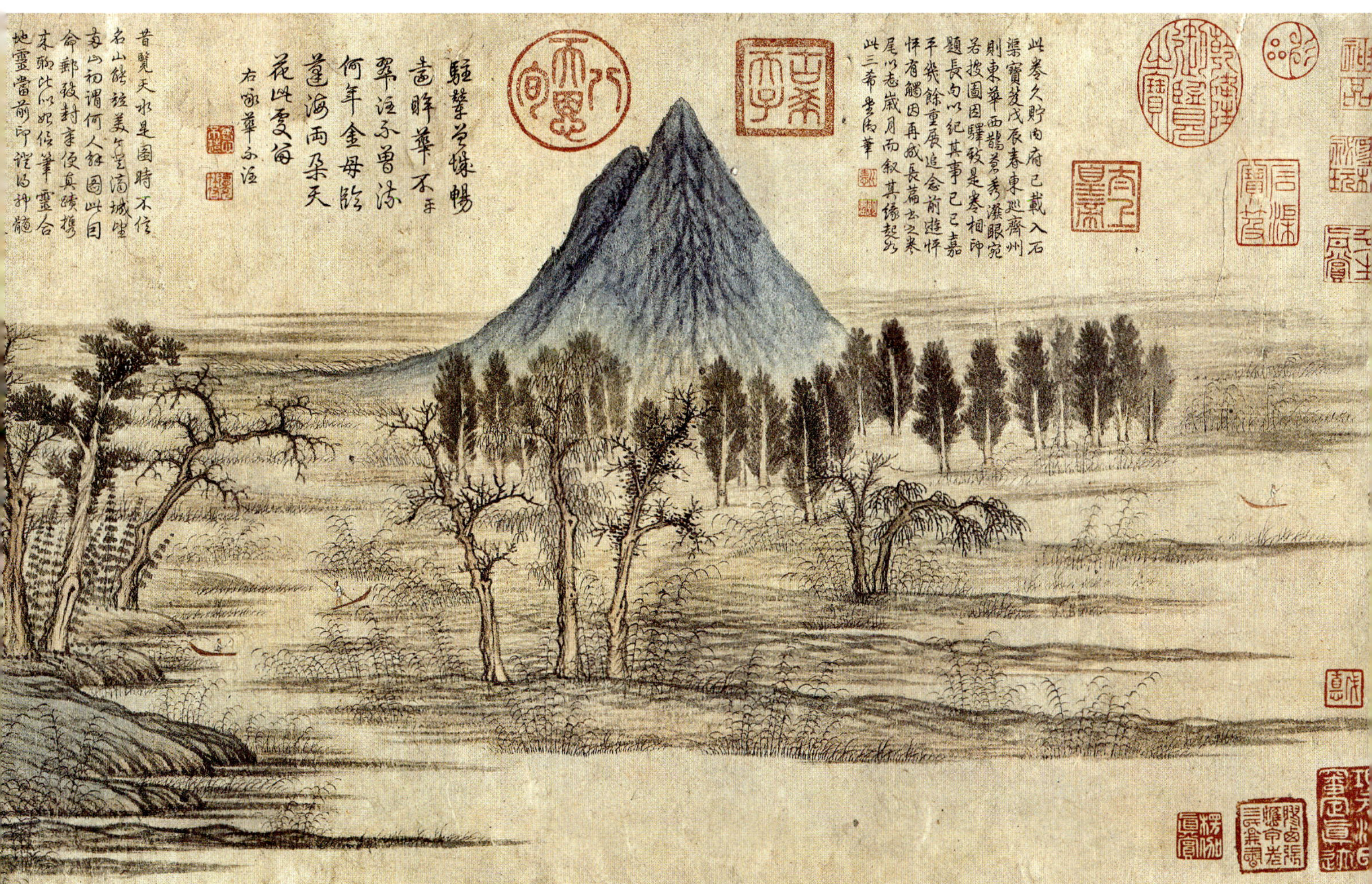

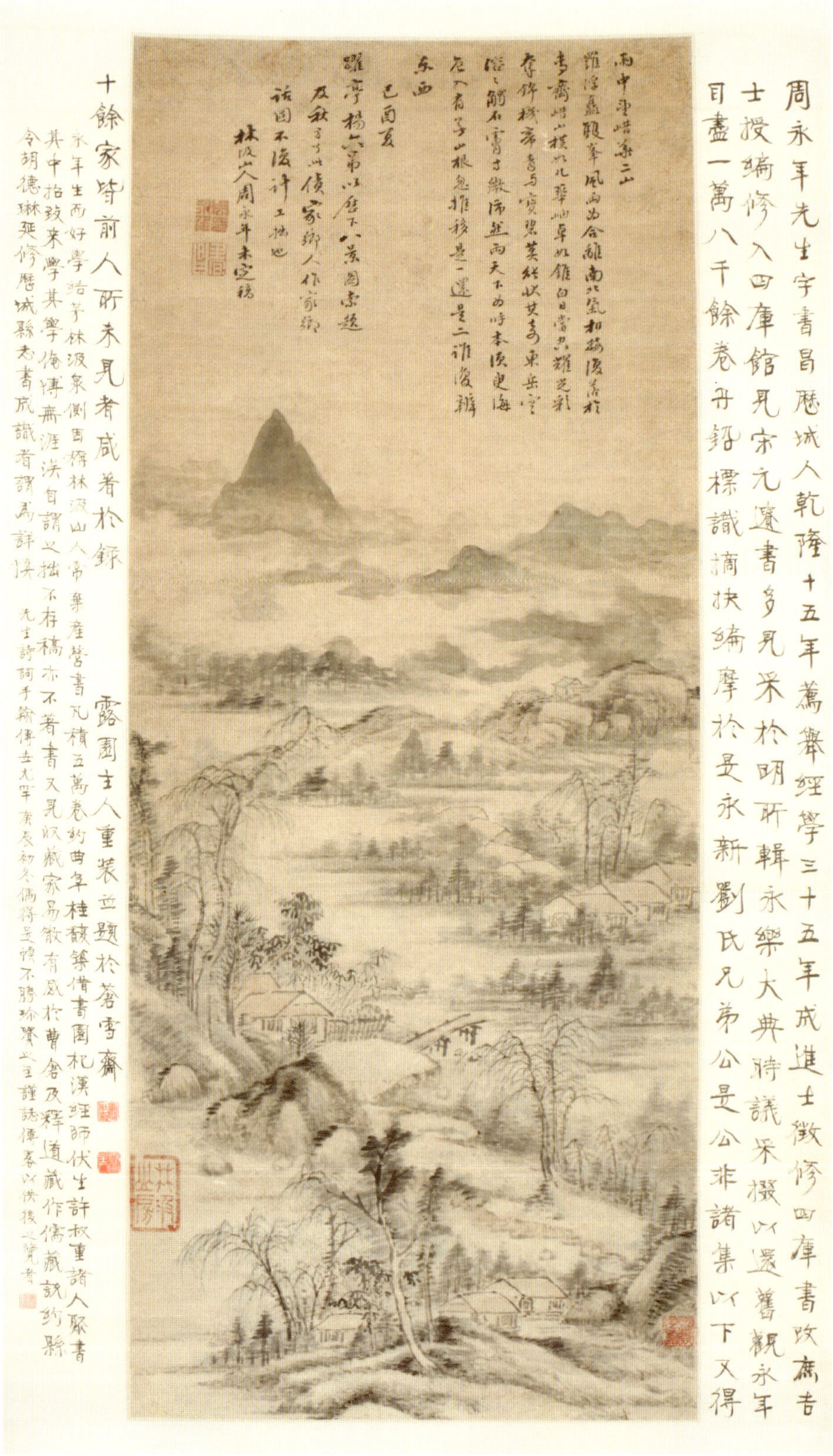

周永年题烟雨鹊华图轴

Scroll Painting of "Misty Rain over Que and Hua Mountains" with Inscription by Zhou Yongnian

清（1644—1911）
山东博物馆藏

周永年，字书昌，历城（今山东济南）人，乾隆年进士，学者，藏书家。少而好学，喜购书，积五万余卷，与曲阜桂馥筑“藉书园”，便于学者览阅、传抄。曾负责《永乐大典》的辑校工作，著有《儒藏说》，曾参与编纂《历城县志》《东昌府志》。

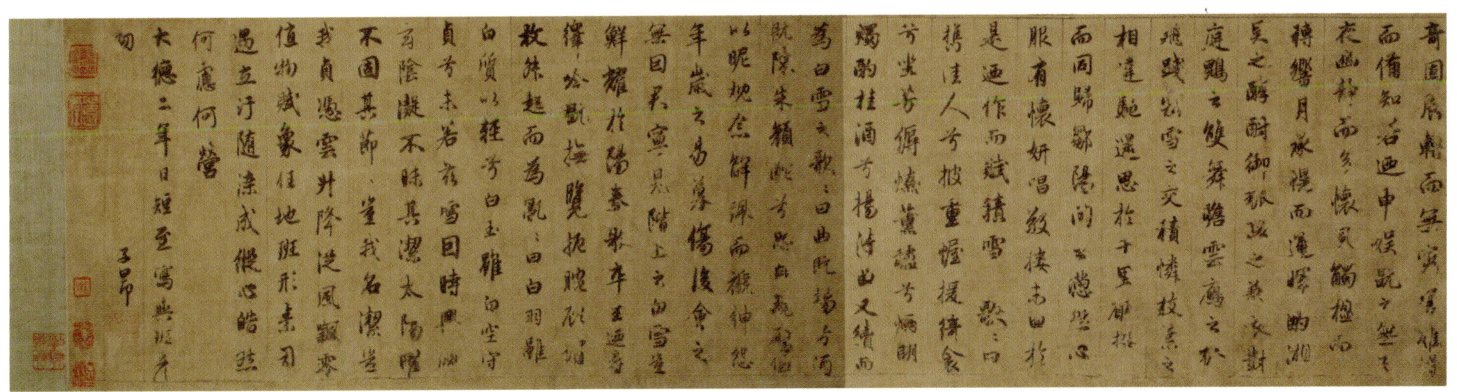

赵孟頫雪赋卷

Zhao Mengfu's Snow Rhapsody Scroll

元大德二年（1298）
山东博物馆藏

《雪赋卷》为元大德二年赵孟頫为班惟志所书，是研究赵孟頫及其书学的重要作品。

蒲松龄：年年作客芰菱乡

蒲松龄与济南的关系十分密切，用他自己的话说是“年年作客芰（jì）菱乡”。他多次游览大明湖、趵突泉、白雪楼等名胜古迹，流连其间不禁为之心醉，留有诸多诗赋赞咏济南的明湖秋月、佛山菊黄。其中，他对趵突泉的喜爱，可从其所著500余言的《趵突泉赋》中可见一斑。

Pu Songling: A Wanderer in the Realm of Jinan Every Year

Pu Songling shared an intimate bond with Jinan. In his own words, he was “a wanderer in the realm of Jinan every year” . He visited renowned scenic spots such as Daming Lake,Baotu Spring,and White Snow Tower for many times,lingering among them and becoming utterly enraptured by their charm.He composed numerous poems and verses praising Jinan's autumn moon over the Daming Lake and the golden chrysanthemums blanketing Thousand Buddha Mountain.His fondness of Baotu Spring is notably reflected in his 500-word prose poem Ode to Baotu Spring.

蒲松龄（1640—1715），字留仙，一字剑臣，号柳泉居士，山东省临淄县（今淄博市淄川区）人，清代杰出文学家。少时即有才名，但屡试不中，71岁才成为贡生。能诗文，善作俚曲，著有《聊斋志异》等。

Pu Songling (1640–1715), courtesy name Liuxian or Jianchen, literary name Liuquan Jushi (Willow Spring Hermit), was born in Linzi County, Shandong Province (now Zichuan District, Zibo City). He was an eminent literary figure of the Qing Dynasty. Despite showing literary talent from a young age, he repeatedly failed imperial examinations and only became a tribute student (Gong Sheng) at the age of 71. He was proficient in poetry and prose, excelled at composing vernacular songs, and authored literary works including *Strange Tales from Liao Zhai* (*Liaozhai Zhiyi*).

济南八景：趵突泉涌

天下第一泉——趵突泉是泉城济南的一大奇景，至迟在明代开始，第一泉就已经成为趵突泉的别名了。水盛之时，泉水喷涌，上冲亭歌，雄豪绮丽，气象万千。主泉四周多小泉，有的像鲤鱼吐泡，有的像绿色丝线穿起的串串珍珠，还有的像绽开的珠花，把三窟的瀑流衬托得更加气势壮观。趵突泉不论是白天还是夜间，是春夏还是秋冬，都呈现出不同的景观。夜间，天宇澄清，月光明澈，波浪鼓荡，满池的“碎玉”，闪闪发光。冬天满池的水荇，纵横缭绕，翠色盈裳，涌出柔润的水气。

Gurgling Baotu Spring, One of Eight Scenic Spots of Jinan

Baotu Spring, the first spring in the world, is a great wonder in the spring city of Jinan. Since the Ming Dynasty at the latest, the first spring has become the alias of Baotu Spring. At the peak of its flow, the spring surges with such vigor that its waters leap skyward like a roaring anthem, presenting a spectacle of majestic grandeur and boundless dynamism. Surrounding the main spring, clusters of many small springs emerge in varied forms, some resembling carps spitting bubbles, some like strings of pearls threaded by green silk, and still others mimicking blooming crystal flowers. Their delicate interplay sets off the waterfalls of the three caves even more magnificent. Baotu Spring presents different landscapes whether it is day or night, spring, summer or autumn. Under the pristine and boundless night sky bathed in crystalline moonlight, the pool churns with waves, and its surface shimmers like a scattering of “shattered jade”. In winter, the pond brims with aquatic tendrils, and their sinuous forms interlaces beneath icy waters, and is cloaked in jade-green radiance, exhaling a veil of mist so delicate it seems as if woven from liquid silk.

趵突泉历史影像
Historical images of Baotu Spring

四时风物

泉水孕育了济南独特的四时风物，使得这座城市四季皆景，还塑造了独具特色的泉水生活传统、寄情泉水的文化审美与表达。泉水已成为济南人生活中不可或缺的一部分，城依泉神韵信增，泉依城源远流长。泉的清纯灵秀，湖的包容开放，山的浑厚俊秀，河的清逸优雅，全在济南这座城市呈现出来。泉是自然禀赋，上天所赐，钟灵毓秀；泉文化是精神创造，人文滋养，熠熠生辉。

春和景明

曲水流觞

曲水流觞起源于古老的祓禊（fú xì）祭祀活动。每年上巳日（夏历三月初三）人们在水边举行祭礼，洗濯去垢，消除不祥。自魏晋时期以来，上巳修禊逐渐演变成一种文人风雅之事。士大夫们在河渠两旁集会宴饮，在上流放置酒杯，使之顺流而下，停在谁的面前，谁就取杯饮酒，称“曲水流觞”。北魏郦道元在《水经注》中明确记载了当时的济南依托泉溪“引水为流杯池”的宴饮活动。济南是北方唯一利用泉水并以较大规模公共空间进行曲水流觞活动的城市。

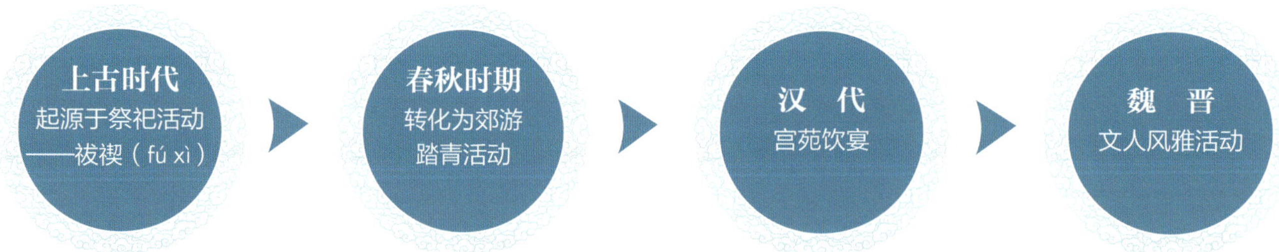

“曲水流觞”发展演变示意图
Schematic Diagram of the Development and Evolution of “Qu Shui Liu Shang” (Drinking Water from a Winding Stream with One Wine Cup Floating on it so as to Wash Away Ominousness)

人物纹双鹿耳玉杯
Jade Cup with Two Dear-Pattern Handles

清乾隆（1736—1796）
济南市博物馆藏
高 5.5 厘米，口径 8.2 厘米

白玉质。玉杯口沿至圈足处置双耳，双耳上方各圆雕一卧鹿，口衔灵芝。

Scenery of Four Seasons

The spring water has nurtured the unique four-season scenery of Jinan, making the city a scenic spot throughout the year. It has also shaped a unique tradition of spring water life, cultural aesthetics and expression of longing for spring water. Spring water has become an indispensable part of the lives of Jinan people. The charm of spring water in the city has increased, and spring water has a long history in the city. The purity and elegance of springs, the inclusiveness and openness of lakes, the richness and handsomeness of mountains, and the tranquility and elegance of rivers are all presented in the city of Jinan. Springs are a natural endowment, bestowed by heaven, with exquisite and graceful spirits. Spring culture is a spiritual creation, nourished by the humanities, and shines brightly.

Gentle Warm Spring and the Bright Scenery

Qu Shui Liu Shang

The tradition of "Qu Shui Liu Shang" originated from the ancient Fuxi purification rites, a ritual of ceremonial cleansing by water. Every year on the last day (the third day of March in the summer calendar), people held rituals by the water to wash and descale to eliminate ominous. Since the the Wei and Jin dynasties, the Spring Purification Ritual gradually evolved into an elegant literati tradition, merging ritual with poetic refinement. The literati gathered on both sides of the river to feast, placing wine glasses upstream and letting them flow downstream. Whoever stopped in front of them would take a glass and drink, known as "Qu Shui Liu Shang". Li Daoyuan of the Northern Wei Dynasty explicitly documented in Commentary on the Waterways Classic how Jinan, leveraging its natural springs and streams, "channeled water to create a floating cup pond" for poetic wine banquets. Jinan is the only city in Northern China that uses spring water and uses large-scale public spaces for activities of "Qu Shui Liu Shang".

泉水烹茶

唐代，受到禅宗品茗助禅的影响，煎茶饮茶文化在济南地区兴起。宋代品茗鉴水之风盛行，由于济南泉水“清、醇、甘、冽”的特性，使得济南泉茶为名士所钟爱，以名泉烹茶、试茶成为一道风雅之事。在灵岩寺甘露泉上方岩壁上，刻有“政和五年（1115）七月，季德修五人就甘露泉试北苑茶”的字样。饮茶风俗随商贾与僧侣传入济南古城，泉水周边遂出现大量饮茶之所。至今，济南人仍有汲泉水泡茶的习惯。

Tea-boiling with Spring Water

In the Tang Dynasty, influenced by Zen Buddhism's practice of drinking tea to aid Zen, the culture of making and drinking tea emerged in the Jinan area. During the Song Dynasty, the practice of tea connoisseurship and water appraisal flourished. Jinan's spring water, celebrated for its clarity, mellowness, sweetness, and crispness, made Jinan Spring Tea highly esteemed among literati elites. Brewing and tasting tea with these famed springs evolved into a refined ritual of scholarly elegance. Carved on the cliff above Ganlu Spring at Lingyan Temple is the inscription: "In July of the fifth year of the Zhenghe era, Ji Dexiu and four others tried Beiyuan Tea at Ganlu Spring." The custom of drinking tea was introduced to the ancient city of Jinan with merchants and monks, and a large number of tea drinking places appeared around the spring water. To this day, Jinan people still have the habit of drawing spring water to make tea.

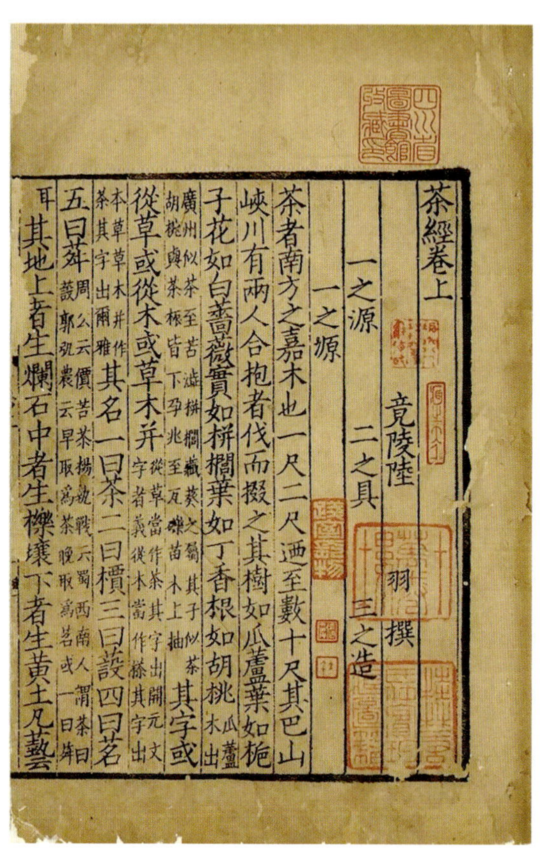
茶經卷上
竟陵陸羽撰
一之源　二之具　三之造
一之源
茶者南方之嘉木也一尺二尺迺至數十尺其巴山峽川有兩人合抱者伐而掇之其樹如瓜蘆葉如梔子花如白薔薇實如栟櫚葉如丁香根如胡桃瓜蘆木出廣州似茶至苦澀栟櫚蒲葵之屬其子似茶胡桃與茶根皆下孕兆至瓦礫苗木上抽其字或從草或從木或草木并從草當作茶其字出開元文字音義從木當作梌其字出本草草木并作荼其字出爾雅其名一曰茶二曰檟三曰蔎四曰茗五曰荈周公云檟苦荼揚執戟云蜀西南人謂茶曰蔎郭弘農云早取爲荼晚取爲茗或一曰荈耳其地上者生爛石中者生櫟壤下者生黃土凡藝

《茶经》书影
Printed Matter of *the Classic on Tea*

《茶经》谈到宜茶用水：“其水，用山水上，江水中，井水下。其山水，拣乳泉、石池漫流者上。”意为经过砂石过滤漫流出的泉水为烹茶佳品。

趵突泉畔望鹤亭茶舍
the Tea House of Wanghe Pavilion by Baotu Spring

宜兴窑桃形杯

Peach-shaped Cup

清（1644—1911）
济南市博物馆藏
杯口长 13.2 厘米，杯口宽 9.5 厘米

杯为连体双桃，一桃完整，另一桃切开为半。栗黄色，胎质细腻，造型自然，别有情趣。

济南八景：锦屏春晓

龙洞山在济南市东南 30 里处。这里山势奇绝，风景壮丽。作为济南八景之首的“锦屏春晓”就在风景最佳的“独秀峰”。明代文人刘敕曾做过这样的描绘：“丹碧点缀，晓霞掩映，绚若锦屏。”西峰悬崖上是西龙洞，洞壁镌有佛像，洞顶钟乳石花丛生。东龙洞在东悬崖上，难以攀登，相传有“金瓶”“春晓”两洞，每逢立春这天，有阳气冲出，干叶、枯草随之飞扬。龙洞山下的深谷中，原建有“寿圣院”，形式古拙，金碧辉煌，院内松柏苍翠，古碑数方。涧内有泉，流水潺潺，形成山清水秀的景观。二三月间，桃花迷径，旭日东升，景色尤奇，一派生机。

Spring Dawn at Brocade Screen, One of the Eight Scenic Spots of Jinan

Longdong Mountain is 30 li southeast of Jinan City. The mountains here are stunning and the scenery is magnificent. As the premier scenic wonder among the Eight Scenic Spots of Jinan, “Spring Dawn at Brocade Screen” is located at Duxiu Peak, an area celebrated for its unparalleled beauty. The Ming scholar Liu Chi once depicted it like this:“Vermilion cliffs and jade-hued foliage dapple the scene; dawn’s rosy clouds interplay with their contours, weaving splendor akin to a brocade tapestry.” On the cliff of Xifeng is Xilong Cave, with Buddha statues engraved on the cave walls and stalactites blooming on the cave roof. Donglong Cave is located on the eastern cliff and is difficult to climb. Legend has it that there are two caves, “Jinping” and “Chunxiao”. On the day of the beginning of every spring, there is a surge of yang energy, and dry leaves and withered grass fly with it. In the deep valley under the Dragon Cave, there used to be the “Shousheng Courtyard”, which was ancient and magnificent in form. There is a spring in the stream, and the flowing water is gurgling.In the second and third lunar months, peach blossoms cloak the winding paths in a haze of pink, while the morning sun ascends, casting its golden light upon a landscape of extraordinary marvel.

采莲消夏

据唐代段成式《酉阳杂俎》记载，曹魏时，大明湖荷花盛开之际，一些官吏、文人到湖边避暑，他们把荷叶割下，盛上美酒，然后用簪子将荷叶中心刺开，使之与空心的荷茎相通，然后从荷茎末端啜饮。这就是被唐宋文士传为美谈的“碧筒饮”。“碧筒饮”深得白居易、戴叔伦、苏轼等名士推崇，逐渐传到湖南、浙江、江西等地，成为一种影响广泛的文化风俗。

Picking Lotus Leaf for Summer

As recorded in Duan Chengshi's *Miscellaneous Morsels from Youyang* of the Tang dynasty, during the Cao Wei period, when lotus flowers bloomed in full glory at Daming Lake, officials and literati would gather by the lakeshore to escape summer's heat. They ingeniously cut lotus leaves to serve as wine vessels, filled them with fine wine, then pierced the leaf's center with hairpins, aligning the perforation with the hollow stem. Finally, they sipped the wine through the stem's end in a ritual blending botanical ingenuity and poetic indulgence. This is the "Bi Tong Drinking" that was passed down as a beautiful talk by literati of Tang and Song Dynasties. The practice of "Bi Tong Drinking" was highly praised by famous scholars such as Bai Juyi, Dai Shulun and Su Shi, and gradually spread to Hunan, Zhejiang, Jiangxi and other places, becoming a widely influential cultural custom.

金廷标《莲塘纳凉图》
（上海博物馆藏）
Painting of Cool-Enjoying in the Lotus Pond by Jin Tingbiao(Collected by Shanghai Museum)

夏日大明湖（王琴/摄）
The Daming Lake in Summer

青花束莲纹盘

Blue-and-white Porcelain Plate with Lotus Spray Design

明宣德（1426—1435）
济南市博物馆藏
高 7.4 厘米，口径 40.5 厘米，底径 29.9 厘米

直口，浅弧形壁，矮圈足。通体施白釉青花纹饰。

荷叶式玉洗

Lotus leaf-style Jade Water Vessel for Washing Writing Brush

明（1368—1644）
济南市博物馆藏
长 15.5 厘米，宽 9.46 厘米，高 4.1 厘米

由一整块玉石挖空雕刻而成，整体为一上托荷叶状。玉洗是文房用具之一，一般放置案头用来盛水洗笔。

石雕荷叶洗

Stone-carved Lotus-leaf Water Vessel for Washing Writing Brush

清乾隆（1736—1796）
济南市博物馆藏
高 10.5 厘米，口径 23 厘米

石洗随形，洗面内凹，表面细腻光亮，洗内侧浅浮雕水草一枝。

诗意秋光

清末小说家刘鹗在《老残游记》中描绘了济南的晚秋景色，如“一路秋山红叶，老圃黄花，颇不寂寞”。他细致地描述了趵突泉、金线、黑虎诸名泉当年的情状，并写下了对济南的印象：“家家泉水，户户垂杨，比那江南风景，觉得更为有趣……”从此，“家家泉水，户户垂杨”成了形容济南景色最能传神达意的妙词绝句。

Poetic Autumn Light

At the end of the Qing Dynasty, the novelist Liu E depicted the late autumn scenery of Jinan in *The Travels of Lao Ts'an,* such as “Along the path, autumn mountains blazed with crimson foliage; aged gardens glowed with golden chrysanthemums—such vistas rendered the journey anything but solitary”. He described in detail the conditions of famous springs such as Baotu Spring, Jinxian Spring and Black Tiger Spring in ancient times, and wrote down his impression of Jinan: “Every household has its own spring water, and every household has its own weeping poplar trees. It is more interesting than the scenery of Jiangnan...” From then on, “Every household has its own spring water, and every household has its own weeping poplar trees” became the most vivid and expressive quatrain to describe the scenery of Jinan.

家家泉水，户户垂杨，
比那江南风景，
觉得更为有趣……

刘鹗（1857—1909），祖籍江苏丹徒，清末小说家。他涉猎众多领域，著述颇丰，所著《老残游记》是中国十大古典白话长篇小说之一，也是中国四大讽刺小说之一。

Liu E (1857-1909) , a native of Dantu, Jiangsu, was a novelist at the end of the Qing Dynasty. He has dabbled in numerous fields and written extensively. His book of *the Travels of Lao Ts'an* is one of the top ten classical vernacular novels in China and also one of the four major satirical novels in China.

大明湖铁公祠内景
Interior of Tiegong Temple of Daming Lake

济南八景：佛山赏菊

千佛山东南不远处就是佛慧山，峰峦突兀，涧谷萦回。山腰原建有古刹开元寺，寺周悬崖凿有石室，昔日儒生多在里面读书。殿前崖下有甘露泉，水声叮咚富有琴韵。稍上是望湖亭，于内可观明湖碧波。一到秋天，红叶满山，黄花遍地，风景优美，被称为"佛山赏菊"。明朝人边贡在《九日登佛山》诗中曾有"背领丹枫直，垂岩紫菊肥"的赞语。据志书记载，千佛山南侧原有赏菊崖，是古人赏菊之处。

Appreciating Chrysanthemums at Qianfo Mountain, One of the Eight Scenic Spots of Jinan

Not far from the southeast of Qianfo Mountain is Fohui Mountain, with abrupt peaks and winding valleys. Kaiyuan Temple, an ancient temple, was originally built on the mountainside. There is a stone chamber carved on the cliff around the temple, where many Confucian scholars used to study in the past. There is Ganlu Spring under the cliff in front of the temple, with a sound of tinkling and a rich qin(Seven-string Chinese Zither) melody. Slightly above is Wanghu Pavilion, where you can see the blue waves of Minghu Lake. In autumn, the mountains are covered in red leaves and yellow flowers are everywhere. It is called "appreciating chrysanthemums in Foshan Mountain". In the poem of *Climbing Qianfo Mountain in the Double Ninth Festival*, Bian Gong, a native of the Ming Dynasty, once praised "Scarlet maples stand tall along the ridged spine,while purple chrysanthemums cling lush to the cliff's incline." According to historical records, there was originally a "Chrysanthemum Appreciation Cliff" on the south side of Qianfo Mountain, which was a place for the ancients to appreciate chrysanthemums.

响晴冬日

《济南的冬天》是一篇充满诗情画意的散文，此文一反老舍以往厚重、富有沧桑感的现实主义风格，以轻快、自然的笔调和温情的意象，描绘了济南这块冬天里的宝地，赋予了济南这座古城独特的人文情感和生命活力，表现了作者对济南的热爱与怀念。

Very Sunny and Bright Winter Days

Winter in Jinan is a poetic and picturesque prose that breaks away from Lao She's heavy and weathered realism style. With a light and natural tone and warm imagery, it depicts the precious land of Jinan in winter, endowing this ancient city with unique humanistic emotions and vitality, and expressing the author's love and nostalgia for Jinan.

上帝把夏天的艺术赐给瑞士，
把春天的赐给西湖，
秋和冬的全赐给了济南。

济南八景：白云雪霁

珍珠泉大院原是历代王府官邸，内有濯缨湖，由珍珠、濯缨、珠沙等泉汇集而成。昔日湖面宽阔，景色秀丽，《珍珠泉铭序》中曾描述过："潆泓冲融，清澜百步，旁流带垣，通舟二里；鱼鸟荇藻，怡怡悦性。"岸边建有白云楼，是元朝张宏官邸，元散曲家张养浩曾作《白云赋》赞许："翼截华鹊之烟雨，背摩霄汉之日星。"站在楼上，北可观明湖碧波，黄河帆影，南可望梵宇簇立，群山青葱。"且当雪霁白云缭绕，下接水光，上浮天际，宫殿隐隐在烟雾中，宛然如画。"雪后凭栏远望，晴光四野，景色绮丽，悦人心目。故被人称为"白云雪霁"。

White Cloud Snow Radiance, One of the Eight Scenic Spots of Jinan

The Pearl Spring Courtyard was originally the official residence of the royal palaces of the past dynasties. There is Zhuoying Lake in it, which is formed by the collection of Pearl, Zhuoying, Zhusha and other springs. In the past, the lake stretched wide with sublime beauty. As described in *Preface to the Inscription on Pearl Spring*: "Eddying depths converge in liquid harmony, crystalline waves stretching a hundred paces; side channels embrace walls like liquid silk, navigable waterways spanning two li. Fish and waterfowl weave through swaying macrophytes, and their harmonious dance nourishes the soul." On the shore stands the White Cloud Tower, built as the official residence of Zhang Hong during the Yuan Dynasty. Zhang Yanghao, the renowned Yuan-Dynasty non-dramatic songs writer, immortalized it in *his Ode to White Clouds*: "Its wings shear through Hua-Que mists and rains; its back grazes celestial suns and stars." Standing upstairs, one can see the clear waves of Minghu Lake and the shadows of Yellow River sails to the north, and the clusters of Buddhist temples and lush mountains to the south. "And when the snow falls and white clouds swirl around, with water shining below and floating in the sky, the palace is faintly shrouded in smoke, resembling a painting." After the snow, leaning against the railing and looking into the distance, the clear light spreads across the fields, the scenery is beautiful and pleasing to the heart. Therefore, it is called "White Cloud Snow Radiance".

珍珠泉历史影像
Historical Images of Pearl Spring

参考文献

[1] 山东大学考古学院，山东省文物考古研究院，济南市考古研究院 . 济南市大辛庄遗址 F61 建筑基址发掘简报 [J]. 考古，2024（11）：50—63

[2] 山东大学考古学与博物馆学系，济南市章丘区城子崖遗址博物馆 . 济南市章丘区焦家新石器时代遗址 [J]. 考古，2018（07）：28—43

[3] 山东大学考古系 . 山东长清县仙人台周代墓地 [J]. 考古，1998（09）：11—25

[4] 山东省文物考古研究所，章丘市城子崖博物馆 . 章丘市西河遗址 2008 年考古发掘报告 [J]. 海岱考古，2012（01）：67—138

[5] 山东省文物考古研究所，章丘市博物馆 . 山东章丘市小荆山后李文化环壕聚落勘探报告 [J]. 华夏考古，2003（03）：3—11

[6] 山东省文物考古研究所 . 山东济阳刘台子西周六号墓清理报告 [J]. 文物，1996（12）：4—25

[7] 山东省文物考古研究院，北京大学考古文博学院，济南市考古研究所 . 济南市章丘区东平陵城遗址铸造区 2009 年发掘简报 [J]. 考古，2019（11）：49—66

[8] 济南市文化局文物处，长清县文物管理所 . 山东长清县明德王墓群发掘简报 [J]. 考古学集刊，1997（01）：221—241

[9] 济南市博物馆 . 试谈济南无影山出土的西汉乐舞、杂技、宴饮陶俑 [J]. 文物，1972（05）：19—24.

[10] 济南市考古研究所 . 济南老城区卫巷遗址出土的宋代金银器窖藏 [J]. 中国国家博物馆馆刊，2016（06）：41—53

[11] 济南市考古研究所 . 济南市魏家庄汉代墓葬发掘报告 [J]. 海岱考古，2015（01）：164—367

[12] 济南市考古研究院，山东大学考古学与博物馆学系，山东省文物考古研究院 . 济南市大辛庄遗址商代墓葬 M235、M275 发掘简报 [J]. 考古，2021（09）：12—23

[13] 平阴县博物馆筹建处 . 山东平阴洪范商墓清理简报 [J]. 文物，1992（04）：93—96

[14] 张溯 . 济南章丘西河后李文化聚落的初步分析 [J]. 东南文化，2020（01）：75—84

[15] 徐慧敏 . 明代德王府对济南古城空间格局演变的影响研究 [D]. 中国建筑设计研究院，2019

[16] 李森 . 龙兴寺历史与窖藏佛教造像研究 [D]. 山东大学，2005

[17] 刘丽丽，郝素梅，柴懿 . 山东济南出土的北齐石俑 [J]. 文物天地，2023（01）：104—108

[18] 孙华昱 . 济南地区墓志研究 [D]. 曲阜师范大学，2023

[19] 李曰训 . 山东章丘女郎山战国墓出土乐舞陶俑及有关问题 [J]. 文物，1993（03）：1—6

[20] 石风 . 出土墓志所见北朝女性世界 [D]. 山东师范大学，2024

后　记

“一城山色半城湖——济南泉·城文化景观特展”由中国文化遗产研究院、山东省文化和旅游厅指导，济南市文化和旅游局、山东博物馆主办，济南市博物馆、济南市考古研究院承办。本次展览选取13家参展单位500余件代表性文物，汇集济南文物精华，以济南古城的历史变迁、泉水文化、人文风情和申遗成果为基本故事线，向观众讲述“济南泉·城文化景观”的千年故事，展现泉城济南的古城格局、自然历史风貌和首府地位，推动古城与申遗工作共同发展。

展览为首次在省级博物馆全面展示济南的历史沿革和文物，策展和文物征集工作都面临较大的困难，济南市文化和旅游局积极组织协调，山东博物馆在场地提供、设备支持、专家指导、展览实施落地等各个环节都给予了大力的支持和帮助，各相关单位克服重重困难，通力合作，使筹展工作得以顺利进行。济南市博物馆、济南市考古研究院为展览的策划实施付出了辛勤劳动，“济南泉·城文化景观”申遗部分的内容由北京清华同衡规划设计研究院有限公司做了大量审修工作，在展览筹备过程中还得到了山东大学博物馆、山东大学考古学院、山东省文物考古研究院、山东省图书馆、章丘区博物馆、长清区博物馆、历城区博物馆、济阳区博物馆、平阴县博物馆、城子崖遗址博物馆的大力支持，在此特向他们表示诚挚的谢意。

限于学识水平和编写时间，本书难免存在不尽如人意之处，错误也在所难免，敬请读者批评指正。

编者

2025年5月

Postscript

"A City With Green Mountains and Half A City Surrounded by Lakes: A Special Exhibition on Jinan Spring-City Cultural Landscape" is organized under the guidance of the China Academy of Cultural Heritage and the Shandong Provincial Department of Culture and Tourism, hosted by the Jinan Municipal Bureau of Culture and Tourism and Shandong Museum, organized by Jinan Museum and Jinan Institute of Archaeology. The exhibition features over 500 representative cultural sets from 13 participating institutions, showcasing the essence of Jinan's cultural heritage. Centered around the historical evolution of Jinan's ancient city, spring water culture, humanistic traditions, and achievements in World Heritage nomination, it narrates the millennial story of "Jinan Springs-City Cultural Landscape". The exhibition demonstrates Jinan's urban layout as the Spring City, its natural and historical features, and its status as a provincial capital, promoting the coordinated development of ancient city preservation and World Heritage application.

This exhibition is the first time to comprehensively display the historical evolution and cultural relics of Jinan in a provincial museum, both the curation and the collection of cultural relics have faced considerable difficulties. Through active coordination by the Jinan Municipal Bureau of Culture and Tourism, Shandong Museum provided substantial support and assistance in various aspects such as venue provision, equipment support, expert guidance, and the implementation and execution of the exhibition, all participating institutions overcame difficulties with collaborative efforts to ensure smooth progress. Jinan Museum and Jinan Institute of Archaeology dedicated extensive efforts to refining the exhibition framework, while the World Heritage nomination section of "Jinan Spring-City Cultural Landscape" benefited from rigorous revisions by Beijing Tsinghua Tongheng Urban Planning and Design Institute. During the preparation of the exhibition, it received strong support from the Shandong University Museum, School of Archaeology at Shandong University, Shandong Institute of Cultural Relics and Archaeology, Shandong Library, Zhangqiu District Museum, Changqing District Museum, Licheng District Museum, Jiyang District Museum, Pingyin County Museum, Chengziya Site Museum. We hereby express our heartfelt gratitude to all contributors.

Given constraints in academic capacity and editing time, this publication may inevitably contain imperfections and inevitable errors. We sincerely welcome criticism and corrections from readers.

Editors

May 2025

图书在版编目（CIP）数据

一城山色半城湖：济南泉·城文化景观特展：汉英对照 / 济南市文化和旅游局编. -- 济南：济南出版社，2025. 4（2025.10 重印）. -- ISBN 978-7-5488-7193-4

Ⅰ. TU-856

中国国家版本馆 CIP 数据核字第 20250M9W43 号

一城山色半城湖：济南泉·城文化景观特展

YI CHENG SHANSE BAN CHENG HU: JINAN QUAN·CHENG WENHUA JINGGUAN TEZHAN

济南市文化和旅游局　编

出 版 人　谢金岭
责任编辑　袁　满　王璐瑶
名士画像　周　群
装帧设计　胡大伟

出版发行　济南出版社
地　　址　山东省济南市二环南路 1 号（250002）
总 编 室　0531-86131715
印　　刷　济南新先锋彩印有限公司
版　　次　2025 年 4 月第 1 版
印　　次　2025 年 10 月第 2 次印刷
开　　本　210mm × 285mm　16 开
印　　张　24.5
字　　数　320 千字
书　　号　ISBN 978-7-5488-7193-4
定　　价　179.00 元

如有印装质量问题　请与出版社出版部联系调换
电话：0531-86131736